Root Hairs
The 'Gills' of Roots
Development, Structure and Functions

Root Hairs
The 'Gills' of Roots
Development, Structure and Functions

V. Bhaskar

Professor of Forestry & Coordinator,
National Afforestation & Eco-Development Board (Regional Centre),
University of Agricultural Sciences,
Bangalore, India

Foreword by
GRAEME P. BERLYN
School of Forestry and Environmental Studies,
Yale University,
USA

Science Publishers, Inc.
Enfield (NH), USA   Plymouth, UK

SCIENCE PUBLISHERS, INC.
Post Office Box 699
Enfield, New Hampshire 03784
United States of America

Internet Site: *http://www.scipub.net*

sales@scipub.net (marketing department)
editor@scipub.net (editorial department)
info@scipub.net (for all other enquiries)

```
        Library of Congress Cataloging-in-Publication Data
Bhaskar, V.
Root hairs : the 'gills' of roots, development, structure, and
functions / V. Bhaskar : foreword by Graeme P. Berlyn.
     p. cm.
  Includes bibliographical references (p. ) and index.
  ISBN 1-57808-274-9
     1. Roots (Botany) I. Title.
QK644.B48 2003
575.5'4—dc21                                        2003041514
```

ISBN 1-57808-274-9

© 2003, Copyright reserved

All rights reserved. No part of this publication may be reproduced, stored in a retrieval system, or transmitted in any form or by any means, electronic, mechanical, photocopying or otherwise, without the prior permission from the publisher. The request to produce certain material should include a statement of the purpose and extent of the production.

Published by Science Publishers, Inc., Enfield, NH, USA.
Printed in India.

*Dedicated in
Loving memory to
my father
V. Muddalinganna*

Dr Graeme P. Berlyn

Foreword

In this book 'Root Hairs–The Gills of Roots', Professor Bhaskar provides the reader with a comprehensive treatise on root hairs that will serve as a landmark for further research. Root hairs are known to increase the surface area of roots to the point where they can comprise a great deal of the entire absorptive surface of root systems. Root hairs do this at rather minimal carbon cost and because of their range of sizes, they can penetrate the smaller soil pores that are unavailable to the main root systems. Ranging from 80 to 1500 μm in length and from 5-17 μm in thickness, these essentially unicellular structures emanate from epidermal cells and have enormous densities of plasmodesmatal connections to the subjacent cortical cells. They are also highly packed with mitochondria. Root hairs have a life span of ca. 2-10 days, being initiated at the point where cell elongation ceases and mature at the point of maturation of the first xylem cells. Thus, they are well suited to their absorptive role, including provision for providing the energy necessary for active uptake of mineral nutrients.

An interesting question posed by Professor Bhaskar is whether the high rates of respiration of root hairs have any additional roles in the function of the rhizosphere. This book presents hypotheses on this subject from a holistic point of view. Dr Bhaskar covers the occurrence, life history, cytology, anatomy, morphology, and physiology of root hairs. He delves into the critical role of root hairs in the development of nodules—both rhizobial and actinorrhizal—in the process of symbiotic nitrogen fixation. Less well known, but still important is the role of root hairs in mycorrhizal symbioses and their importance for the mineral nutrition of plants.

Dr Bhaskar is recognised for his pioneering work on the respiratory role of root hairs and their ecophysiological significance. He has devoted several chapters to this function viewing root hairs as respiratory centres of roots. The consequences of such a role need to be further investigated and fully appreciated for their significance to the biosphere as a whole. This present volume draws attention to the root hair in particular and provides a thorough review of the pertinent literature on the subject. In addition, this work should stimulate future researchers to investigate further the many fascinating attributes of the world below the surface of the earth and its remarkable features and interactions.

11 September 2001

Dr Graeme P. Berlyn
Professor of Anatomy and Physiology of Trees,
School of Forestry and Environmental Studies,
Yale University,
USA

Preface

My interest on root hairs can be traced to 1985, when I was studying the remarkable growth of *Ailanthus malabarica*, an indigenous tree of Western Ghats, south India, which had been introduced on an experimental basis to the much drier belt of interior Karnataka. Out of curiosity, the roots were examined for the possible symbiotic association with ectomycorrhiza. Interestingly enough, most part of roots, including the mature roots, were densely covered with persistent live root hairs.

This discovery led me to continue studies on root hairs while studying at Yale University during 1987-88. As a part of the curriculum fulfillment, I chose to undertake a project work on root hairs with Professor Graeme P. Berlyn, who asked me to gather all the available literature on root hairs. The Kline Library provided the best opportunity to access all the existing literature on this subject. Despite an exhaustive collection of research matter on root hairs, their actual functional role was somehow not particularly convincing. After an in-depth review of the literature on root hairs, including their ontogeny, structure, histochemistry and functions, I was not convinced that their absorptive role was the main function, as believed by a large majority of botanists and physiologists worldwide.

During this period, I came across matured pods of garden pea (*Pisum sativum* L.) with viviparously-germinated seeds covered with dense white root hairs on the radicles. This led me to inquire as to the reason why root hairs should be produced in such a great number in a gaseous environment (air) prevailing inside the pod, where there was neither water nor any liquid nutrient medium to absorb. This resulted in the discovery of the 'breathing role' in roots being the chief function of root hairs. On tallying all the previous results, they were found to be actually providing ample evidences towards their role in gaseous exchange (mostly gaseous oxygen intake) as it happens through the stomata and lenticels in the aerial parts, although the earlier authors had attributed their role to increase the surface area in order to enhance the absorption of water and minerals as their chief function.

I sent a paper on this new theory to the *New Phytologist* for publication. The referees wrote, 'two hundred years of research has been done on root hairs and their absorptive function has been well proved and the author's new hypothesis on their respiratory role is unbelievable'. Later, the *Journal of Sustainable Forestry* in New York published this article in 1993.

Dr Connolly and Dr Berlyn of the School of Forestry and Environmental Studies, Yale University, continued research on root hairs and provided the first experimental evidence during 1996, providing evidence that root hairs are the main respiratory centres in the entire root system. Some of the recent studies by the present author have further provided several additional evidences in favour of the new theory. More precisely, root hairs have been found to be very sensitive to direct contact with water or any other liquid medium.

It seems obvious that nature's creation of root hairs is primarily meant for 'breathing' in roots, which are subterranean and usually lack special openings such as stomata or lenticels, as found in the aerial part of the plant, while the uptake of minerals and incidentally water, depends to a great extent on the energy produced by root hair respiration. This finding warrants research towards promotion of normal root hair development through proper soil and water management to naturally improve crop health and production with minimized inputs. Oxygen enrichment through root hairs seems to have significant implications in agriculture and forestry. There is a lot that we don't know about root physiology, necessitating a lot more research in this field.

I would appreciate receiving views, comments and suggestions, if any, from the readers at my address mentioned below.

4 June 2002
Bangalore

Dr. V. Bhaskar
Professor of Forestry & Coordinator Regional Centre,
National Afforestation & Eco-Development Board,
(Min. of Envt & Forests, GOI)
University of Agricultural Sciences,
GKVK Campus, BANGALORE 560 065
E-mail: bveeralinga@yahoo.co.in (or)
vb1949@tatanova.com

Acknowledgement

This study on root hairs and the new hypothesis on their functional role as 'respiratory gills' in roots would not have been possible without the generous assistance of the Ford Foundation, which sponsored the academic training at the School of Forestry and Environmental Studies, Yale University. I am extremely grateful to Dr John C. Gordon, then Dean of the Yale School of Forestry and Environmental Studies for his encouragement as also providing facilities during this study.

Perhaps words can never fully express my deep and profound sense of gratitude to Dr Graeme P. Berlyn, Professor of Anatomy and Physiology of Trees, Yale School of Forestry and Environmental Studies, who was generous enough to accommodate my research project on root hairs. I am highly indebted to him for his encouragement, patience and insights. I am particularly indebted to the Kline Library, Yale University, for its largest collections of journals and books, which offered me an ideal opportunity to gather all the existing literature on root hairs and related aspects. I am thankful to the Librarian and Sri Vasu, UAS Library, Bangalore for providing CD-ROM facility for accessing the latest literature.

I am thankful to all those who patiently listened to my discussions, especially, Sriuths Giridhar Pai, Dr C.G. Kushalappa, Dr Chandrashekhar, Dr Shesha, Dr Nataraj Karaba, Dr V. Suresh Babu,

Sri N.V. Srinivasalu and Murali Raghavendra Rao and also Sriuths N. Veerendra Babu, G. Hanumantha Raju and especially my wife Smt. Usha, H., who rendered assistance in one way or the other. Dr Chandrashekhar and Dr Vasanthakumar Thimakapura have offered their assistance in some of the photographs.

June, 2002
Bangalore,

V. Bhaskar

Contents

Foreword ... vii

Preface ... ix

Acknowledgement ... xiii

Chapter 1 : Introduction ... 1

Chapter 2 : Occurrence and Nature of Root Hairs in Higher Plants ... 5

    1. Vascular plants without root hairs ... 6
    2. Aerial roots and root hairs ... 7
    3. Number and size ... 8
    4. Surface area ... 9
    5. Life span ... 10

Chapter 3 : Structure of Root Hair ... 13

    1. Morphological Structure of root hair ... 13
        a) Branched and unbranched root hairs ... 13
        b) Dimorphic root hairs ... 14
    2. Cytology and ultra-structure of root hair ... 14
    3. Cell Wall of root hair ... 17
    4. Cuticle in root hair ... 18

Chapter 4 : Origin and Development of Root Hair ... 21

1. Ontogeny of root hair ... 21
2. Factors influencing development, differentiation and density of root hairs ... 24
   a) Development of root hairs in aquatic plants ... 24
   b) Role of O_2 in root hair development in terrestrial plants ... 27
   c) Effect of ambient environment on root hair development ... 29
   d) Role of mineral nutrients in root hair development ... 30
   e) Effect of pH on root hairs ... 32
   f) Role of microbes on root hair development ... 32
   g) Role of hormones in root hair development ... 32
3. Genetic control of root hair development ... 33
4. Biotechnology with root hair ... 34

Chapter 5 : Physiological Functions of Root Hairs: Earlier Theories ... 37

1. Role of root hair in water and mineral uptake ... 37
   a) Water absorption ... 38
   b) Ion exchange or root mineral uptake... 39
2. Role of root hair in root nodule formation ... 41
   a) Site of infection on root hair ... 42
   b) 'Nod' factor and root interaction ... 42
   c) Histochemistry of root hair infection in legumes ... 43
   d) Root hair infection in actinorrhizal plants ... 45
3. Root hair and mycorrhiza ... 47
   a) Root hairs: VAM relationship in terrestrial plants ... 47
   b) Root hairs: VAM relationship in aquatic plants ... 48
   c) Root hairs: Ectomycorrhizal relationship ... 50
4. Root hair and root parasitism ... 51
5. Root hair and infection by nematodes and pathogens ... 52
6. Other functions of root hair ... 53

　　　　i) Role of mucigel ... 53
　　　　ii) Role of root hairs in drought resistance ... 53

Chapter 6 : Root Hairs as 'Respiratory Gills': Evidences ... 55

　　　1. Root respiration ... 55
　　　　a) How do roots get oxygen? ... 55
　　　　b) Where does oxygen enter roots? ... 56
　　　2. Root hair as an entry point for oxygen diffusion ... 58
　　　3. Evidences disfavouring water and mineral uptake as the main functions of root hair ... 59
　　　4. Evidences in support of the respiratory role as the chief function of root hair ... 63
　　　　a) Concentration of hairy roots in the soil surface as evidence of their role as 'respiratory gills' ... 63
　　　　b) Absence of root hairs in roots grown in tissue culture medium ... 63
　　　　c) Development of root hairs in soil voids and aerobic environment ... 65
　　　　d) Cytochemical assay prove elevated respiratory activity in root hairs ... 69
　　　　e) Ultrastructural and histochemical evidences ... 71
　　　　f) Soil moisture and root hairs (anaerobic environment and root hairs) ... 74
　　　　g) O_2-dependent water and nutrient uptake by roots (active uptake of water minerals) ... 76
　　　　h) How do aquatic plants respire? (evidences from hydrophytes) ... 77
　　　　i) How do roots of terrestrial plants respire? (evidences from terrestrial plants) ... 82
　　　　j) Stomata in root hair zone ... 83
　　　5. Additional experimental evidences ... 84

Chapter 7 : Implications of the Respiratory Role of Root Hairs in Agriculture and Forestry ... 97

　　　1. Contribution of roots (hairs) to CO_2 evolution ... 97
　　　2. CO_2-O_2 relationship and their effect on root hair respiration, plant growth and yields ... 98

3. Plant responses to soil anaerobiosis ... 100
   a) Morphological adaptations ... 101
   b) Anatomical adaptations ... 103
   c) Biochemical and physiological adaptations ... 105
4. Root hair management for higher yields ... 106

Chapter 8 : Recommendation for future research ... 111

Chapter 9 : Summary ... 115

Chapter 10 : References ... 125

Illustrations ... 169

Index ... **183**

Introduction

All aerobic forms of life require oxygen. An understanding of the oxygen requirement of an organism and the mechanism for its supply are essential factors in understanding its life processes. For air-breathing animals, the essential factors constitute an open passage to the lungs and an adequate concentration of oxygen in the air that they breathe. Most species of fish extract oxygen from the water through their gills. Roots of higher plants, such as the aerial shoot are aerobes and depend upon a supply of the available oxygen molecules from soil environment in order to support respiration and various other energy-dependent metabolic processes. Shoots are endowed with numerous stomata and lenticels interconnected with intercellular spaces that facilitate the entry and distribution of atmospheric oxygen. A plant's own photosynthesis in the green shoot can also supplement some of its oxygen requirements during daytime. But the subterranean parts like roots suffer most frequently from oxygen shortage, mainly due to the fact that: (1) their soil environment is especially prone to water saturation that excludes free available oxygen; (2) CO_2 concentration in soil is usually higher and oxygen is lesser as compared to aboveground atmosphere; (3) more often, soil aeration is poor; and (4) there is greater inability of plants to diffuse free oxygen molecules from the shoot to the root.

Insufficient O_2 causes the plants to undergo anaerobic respiration, fermenting carbohydrates into alcohol and producing only a small amount of energy (ATP). This energy produced is usually insufficient for normal root metabolism, hence many root cells die and decay under short or prolonged flooding of soils, depending upon the species tolerance. Oxygen shortage in roots causes an overall acidification of the cell sap in sensitive plants such as wheat and barley or alkalization of the cytoplasm in resistant plants such as rice. From these perspectives, anaerobic stress in roots is seen to cause severe economic losses in agriculture, horticulture and forestry in many regions of the world. Hence, during the last two to three decades, much research has been initiated to find out the causes of death in anoxic cells, which may be due to self-poisoning by the ethanol formed by alcoholic fermentation, cytoplasmic acidosis and insufficient energy generation.

Plant physiologists have long believed that plant roots respire and grow in a solution agitated with a fresh supply of oxygen. Although a vast amount of information exists on soil-oxygen-plant relationships, it does not completely provide information applicable to plants growing in the soil because the mechanism for oxygen supply is not adequately accounted for. Plant roots do not have their surfaces continuously swept by a solution containing dissolved oxygen. The water films around the respiring surface and in the soil micro-pores are rather stationary. This means that living roots in the soil were believed to obtain their supply of oxygen by the process of diffusion through a moisture film. The diffusion coefficient of oxygen through water or a solution is about 10,000 times slower than diffusion in the gaseous phase. According to Drew (1979), oxygen diffusion in pure water is about three million times slower than it is in air. The limiting factor for oxygen supply is probably diffusion through the moisture film. Therefore, it is necessary to characterize those root parts that have a direct access to gaseous oxygen and aid in root respiration.

Subterranean parts of higher plants have been studied to a lesser extent as compared to aboveground parts and, consequently, their physiological functions are less well understood. Root hairs are a part of the underground system with interesting functions, as they constitute very small structures of the root system. Root hairs (or root trichomes) are slender, thin-walled unicellular tubular evaginations from the outer wall of rhizodermal cells occurring in the 'root hair zone' between the growing root tip and zone of active root elongation on a primary root. Root hairs are indeed an excellent system used for studies on cellular metabolism and experimental cell biology, because their physical position in the epidermis correlates with the developmental and functional stages and also because they can be easily harvested for the molecular and cellular analysis.

One of the earliest experiments utilizing root hairs dates back to middle of the nineteenth century (Ohlert, 1837; Meyen, 1838). Although till date a tremendous literature has accumulated on the structure and function of these tiny microscopic cells, their real functional role in vascular plants has remained intriguing. Various functions have been ascribed to the root hairs. They are still believed to maximize the surface area of the root system in order to increase the absorption of water and nutrient uptake. However, the structural and functional mechanisms by which this is accomplished remain unanswered. The molecular mechanisms of transmembrane uptake of ion influx remains largely unknown. The mucilaginous layer present over the root hairs is presumed to improve the plants water retaining capacity by enhancing the contact or adhesiveness between the root hair and soil particles and also provide a favourable medium for growth of rhizosphere microorganisms. Whether all plant root hairs possess this mucigel and carry out similar function is questionable. When almost all parts in a root can take up active role in water and mineral ion uptake—either in the presence or absence of root hairs—no special absorptive role can be ascribed to root hairs.

No doubt, root hairs have different specialized functions in various plant groups, such as root nodulation in legumes and actinorhizal species, but they do not constitute the primary functions of root hairs in non-nodulating and non-mycorrhizal plants. Thus, it has not yet been possible to draw up conclusion on the actual physiological and adaptive role of root hairs and as such, they remain very interesting cells.

In the present analysis, the author has attempted to cross correlate between the results of various studies conducted so far on root hairs, which has led to synthesis of a new concept on the biological role of root hairs. This new theory explains with abundant evidences the respiratory role of root hairs. Around half of all assimilated photosynthates is translocated from leaves to below-ground organs (Lambers, 1987). Some of these photosynthates are respired to generate metabolic energy for growth, maintenance and transport processes. Does O_2 required for root respiration enter throughout entire root surface, or is there a definite ventilating port for its easy entry into the root? This basic question has not been hitherto answered. Numerous circumstantial evidences support the theory that at least one of the main functions of root hairs is respiration and they are the main sites where respiratory energy is produced in the root system. It would appear that of all factors, root respiration is the most sensitive aspect of plant activity in regard to soil aeration. Since the growth of roots and uptake of water and nutrients are dependent upon energy which is supplied by respiration, it may be assumed that a reduction in the respiratory activity in the root is the first step in the growth-limiting effects of insufficient aeration. Root hairs are known to increase the surface area of the root enormously, approximately 75 to 90% of the surface area of the entire root system, thereby enabling maximum O_2 intake. Thus, root hairs assume greater importance in root respiration and plant growth. Good gas exchange between soil and the ambient atmosphere is essential for maintenance of an appropriate soil atmosphere to promote normal root hair growth. In view of the new understanding on the main role of root hairs as an avenue for intake of gaseous O_2, the author has also discussed various implications in agriculture, horticulture and forest management practices.

Occurrence and Nature of Root Hairs in Higher Plants

Root hairs are generally confined to the root hair zone of the growing tip of the roots (Fahn, 1982). However, in certain plants, there is no distinct root hair zone. For example, in leguminous trees like *Gleditsia*, root hairs occur throughout the length of the taproot (McDougall, 1921) in deep soil as well as in the superficial layers. Similar is the case in plants growing in various types of habitats or throughout the length of second order, lateral roots from the extreme root tip as in *Quercus borealis* (Richardson, 1953) and *Anacardium occidentale* (Bhaskar, present data) or the entire surface of all the roots as in grasses like *Secale sereale* (Dittmer, 1937). In some succulent xerophytes (as *Opuntia*), root hairs occur on the extreme tip of root (Coulter *et al.*, 1911).

Root hair production by *Quercus borealis* is only on second order lateral roots which are scarcely visible to the naked eye and are produced abundantly throughout the length of the roots, and in many cases, from the extreme root tip which lacks a rootcap. Root hairs present on first order laterals are few and short, while no root hairs grow on the main roots (Richardson, 1953). In grasses, root hairs occur in rows and alternate in regular fashion with non-root hair cells (Fahn, 1982). Root hair development has been used as evidence of relationships among genera of Gramineae (Row and

Reeder, 1957). Root hairs in *Arachis hypogea* occur in the form of tufted clusters or rosettes (**Fig. 1c**) only where lateral roots emerge or in the junction of the root axils (Chandler, 1978; Nambiar *et al.*, 1983). Meisner and Karnok (1991) have observed that approximately 66% of the lateral initiates expressed the rosette-type of root hairs. The present author has observed that freshly-induced primary roots in this plant were completely devoid of root hairs.

In tamarind (*Tamarindus indica*), the radicle exhibits very short but densely placed root hairs but the lateral roots which develop very soon in four rows are devoid of root hairs. In *Casuarina equisetifolia*, dense root hairs are distributed throughout the length of slender wire-like radicle (author's own observation) (**Fig. 2b**). Similarly, in case of *Eleusine corocana* seedlings, root hairs are distributed throughout the long and slender roots.

1. VASCULAR PLANTS WITHOUT ROOT HAIRS

Root hairs do not develop in all higher plant taxa nor under all kinds of conditions in any one taxon. None of the Magnoliales among the Angiosperms possesses root hairs (Baylis, 1975; St. John, 1980 and author's own observation in *Polyalthia longifolia* and *Cinnamomum* sp. which belong to Annonaceae and Lauraceae, respectively). Root hairs are also absent in roots associated with ectomycorrhiza (Mexal *et al.*, 1979; Clayton and Bagyaraj, 1984); in Ericales (Brook, 1952) and orchids (Nieuwdorp, 1972) or may be completely suppressed by mycorrhizas as in eucalypts (Chilvers and Pryor, 1965). Root hairs are never abundant in gymnosperms and even then, have been found only on 'long roots'. Root hairs have not been observed at all in Araucariaceae, Taxodiaceae, and Cupressaceae (Romberger *et al.*, 1993).

Aquatic plants: Submerged aquatic plants, with exceptions, are generally devoid of root hairs. Typical examples include, *Eichhornia crassipes*, *Pistia stratiotes*, *Cyperus eleusinoides*, *Myriophyllum triphyllum*, among others (Charlton, 1978). Root hairs are reported to be absent in the mature plants of *Littorella uniflora* and *Lobelia dortmanna* which are grown submerged in sandy sediments (Sondergaard and Laegaard, 1977). This does not confirm the observation by Shannon (1953), who reported the presence of root hairs in *Lobelia dortmanna*. The plants investigated by Shannon were seedlings cultivated in water, however, which according to Sondergaard and Laegaard (1977), may have influenced the development. Root hairs are also reported to occur in halophytes (Uphof, 1962). The root hairs in aquatic plants need a further examination with regard to their exact nature, cellular structure and functional role as compared to the root hairs produced by land plants.

In case of a swamp forest species (*Pterocarpus officinalis*), a majority of the root hairs (and incidentally nodules) were found above the water table, located on large aerial buttresses (Saur *et al.*, 1998). Large trees were found to have modified their environment by accumulating litter between the buttresses, ensuring a certain amount of soil above the water table. Consequently, the root hairs and the nodules were concentrated in a circle 5 m in diameter around the oldest trees. A few nodules (5%) survived below the water table level, provided healthy root hairs were present.

Root hairs are often found only on young seedlings and are totally absent in adult trees, as in the case of root parasites such as *Santalum album* (Rama Rao, 1903; Varadaraja Iyengar, 1965) (**Fig. 2c**). The author has also observed the presence of very few or sparsely scattered root hairs on roots in case of young sandal seedlings (recruits), but a complete absence in older seedlings.

2. AERIAL ROOTS AND ROOT HAIRS

The aerial roots of the date palm and *Pandanus* do not form root hairs, while the adventitious roots of *Kalanchoe fedtschenkoi* produce multicellular hairs in air but stop producing them when the roots penetrate the soil (Popham and Henry, 1955; Cormack, 1962). Whether these multicellular trichomes on aerial roots can be considered as root hairs is doubtful. The outermost cells of velamen (a mono- or multiseriate rhizodermis tissue in most epiphytic and terrestrial Orchidaceae as well as epiphytic Liliaceae and Amaryllidaceae) may bear root hairs (Pridgeon, 1987).

Aerial roots of plants such as *Tinospora cordifolia* of Menespermaceae lack root hairs but are studded profusely with lenticels all along their length. Root hairs require moist air or higher levels of atmospheric humidity to maintain turgidity; otherwise they shrivel. Hence, they are normally not produced on aerial roots.

The coleorhizal epidermis in grasses develop protrusions similar to root hairs and elongate during radicle emergence (Debaene-Gill *et al.*, 1994). The coleorhiza has been considered to be the suppressed primary root (Negbi and Koller, 1962), and hence the term 'coleorhiza hairs', analogous to 'root hairs'.

3. NUMBER AND SIZE

Root hair zone: The root hair zone of most roots is 1 to 4 cm long and is situated just behind the zone of active root elongation (Farr, 1928a; Jaunin, 1988). In *Gleditsia triacanthos*, the root hair zone is about 25 cm long (Farr, 1928c) or even longer, as in *Ailanthus malabarica* (Bhaskar, unpublished data).

Number: Although commonly only about half of the rhizodermal cells produce root hairs, the total number present may be very large. A *Secale cereale* plant, for example, was determined to have more than 14 000 000 000 living root hairs, with as many as 100 000 000 new ones formed each day (Dittmer, 1937).

Size: Root hairs range from 80 to 1500 µm in length (Jensen and Salisbury, 1984; Romberger *et al.*, 1993) and 5 to 20 µ in diameter, depending upon species and cultivars (Dittmer, 1949; Caradus, 1979). Grasses, in general, have comparatively longer root hairs (Dittmer, 1949) and *Arachis hypogea* has unusually wider root hairs with their diameter ranging from 30 to 50 µ (author's own observation) (**Fig. 1c**). Many sclerophyll species of Mediterranean South Africa and

Western Australia often bear remarkably long (up to 2.4 mm) and abundant root hairs (Lamont, 1982). Root hairs may also be unusually long as in some aquatic plants like *Alternanthera senssilis*.

Several environmental factors affecting root hair size include pH (Ewens and Leigh, 1985), concentrations of ions (Tanaka and Woods, 1972; Ewens and Leigh, 1985; Jaunin and Hofer, 1988), water potential (Ekdahl, 1953), physical properties of the soil (Reynolds, 1975; Greenland, 1979), relative humidity (Reid and Bowen, 1979), and rhizosphere microorganisms (Bowen and Rovira, 1976). Within any one species, the hair size is relatively constant, although hairs are longer on roots exposed to moist air than on those growing in the soil or immersed in water (Hesse, 1904). Hairs that arise on secondary roots or on roots of higher order are shorter than those that arise on main roots, although all roots have more or less the same size (Dittmer, 1949; Tanaka and Woods, 1972). The position of the hairs on the root also determines their dimension, which is a matter of age. The shortest hairs are near the distal end of the hair zone, whereas the longest are near the proximal end (Farr, 1928a; Jaunin and Hofer, 1986).

4. SURFACE AREA

The calculated total root hair surface for one rye plant was 407 billion square meter with 14 billion root hairs. In *Secale cereale* the calculated total root hair surface for the entire plant was 4,321.31 square feet (or 400 m^2—nearly twice that of the roots) (Dittmer, 1937). Its total subterranean surface was 22 times that of the transpirational area of the aboveground shoot (Dittmer, 1937). Approximately 75-80% of the surface area of the entire root system of dicots exhibiting secondary thickening is covered with root hairs, although most of the hairs were on roots smaller in diameter, especially on the roots of quaternary division. Root hairs can increase the surface ('effective radius') area of the root enormously—eg. 6x for *Leucodendron*, 26x for *Hypolaena fastigiata* (Lamont, 1982). Taking into account the 2 µm long root hairs, Gullan (1975 cf. Lamont, 1982) proved that the capillaroid rootlets of *Hypolaena fastigiata* accounted for 90% of the surface of its root system.

Variations with regard to length and surface area of root hairs have also been reported in different cultivars of wheat (Gahoonia et al., 1997). Among wheat cultivars, Kraka had the longest and densest root hairs, while those of Kosack were the shortest and sparsest. Root hairs increased the root surface area of Kraka, Foreman and Kosack by 341%, 142% and 95%, respectively. The variation in root hair parameters of the cultivars was related to the quantity of P depleted from the rhizosphere.

Recently, a simple stereological procedure has been demonstrated for systematic, uniformly random sampling of the root system of known length in order to obtain practically unbiased estimates of the total number and dimensions of root hairs (Wulfsohn and Nyengaard, 1999). The numbers and dimensions of root hairs were estimated for five crested wheatgrass (*Agropyron cristatum*) root systems grown for one month in a gel. Less than one hour was required to obtain the estimates of root hair parameters for a single plant. There was low variability of spatial density of root hairs within a given branching order (CV < 15%). However, because of large variation in the total length of laterals, the total number of root hairs varied greatly (CV ~ 70%). On average, root hairs provided half of the total surface area of a root system and a total length 20 times that of the roots.

5. LIFE SPAN

Root hairs are generally ephemeral or transient structures and become worn out after a few hours or days or weeks (Cutter, 1970). However, in certain plants, root hairs may persist for several years, as in tree species such as *Gleditsia triacanthos*, *Gymnocladus dioica* and *Cercis canadensis* (McDougall, 1921), *Ailanthus malabarica*, *Anacardium occidentale* (author's own observation) (**Figs. 3a-c and 4a, 4b**), and some herbaceous plants of Asteraceae (Whitaker, 1923). Such hairs may last for several months and their diameter becomes four times greater than in the early stages of formation (Farr, 1928c).

According to McDougall (1921), the life span of root hairs is as long as the root epidermis. Whitaker (1923) has observed that the presence of persistent root hairs is always correlated with the absence of secondary growth in roots and the secondary growth marks the disappearance of the root hairs. There is no report on whether these persistent root hairs (which are thick-walled—approximately 2 µ, as reported by McDougall or not quite thick-walled, as observed by Whitaker) are living or have lost the living protoplasm. It was not possible to determine whether these root hairs contain living protoplasm for the walls are very thick and dark colored (McDougall, 1921).

Root hairs in *Lepidium sativum* are long and thin-walled under normal conditions, while they become short and their cell wall thickens when grown in sugar solution or insufficient water supply (Palladin, 1918, cf. Whitaker, 1923). McDougall (1921) reports that these persistent root hairs are dark brown. It is not known whether this dark brown colour of the persistent root hairs is due to the thick cell wall or tannin content in the root hairs. As a matter of fact, root hairs in *Equisetum* possess high tannin content and hence, appear dark (Dittmer, 1949). A similar type of dark brown stiff root hairs are also observed by the present author in *Anacardium occidentale* (**Figs. 3a-b**).

Whitaker (1923) observed the root hairs of several members of Asteraceae growing under exact conditions. According to his findings, in *Gleditsia triacanthos*—which exhibited persistent root hairs on the smaller roots—the secondary thickening in the root marked the disappearance of the root hairs. In members of primitive Persoonieae (*Persoonia, Pycnonia* and *Acidonia*) of Proteaceae, feeder roots often bear a dense cover of persistent root hairs up to 6 mm long (Lamont, 1982). Persistent root hairs have also been observed in *Ailanthus malabarica* and *Anacardium occidentale* (author's own observation) (**Figs. 3 and 4**).

CHAPTER 3

Structure of Root Hair

1. MORPHOLOGICAL STRUCTURE OF ROOT HAIR

Root hairs are unicellular, but Popham and Henry (1955) have considered the multicellular hairs produced on the aerial adventitious roots of *Kalanchoe fedtschenkoi* also as root hairs. Dittmer (1949) has examined millions of root hairs on hundreds of plants and has never seen a septate root hair. Hence, the present author feels that the so-called 'multicellular root hair' is a multicellular trichome and not a real root hair. Occasionally, a hair may be found which is peculiarly shaped or even forked but never septate. Chandler (1978) reported that some of the root hairs in *Arachis hypogea* were septate. The present author, after reexamination, found that root hairs in *Arachis hypogea* are all unicellular and none were septate in nature (**Fig. 1c**). So, a root hair is simply a protrusion of the cell from which it arises and always remains just a portion of that particular cell.

a) Branched and unbranched root hairs

The typical root hairs have a cylindrical form developing at a right angle with the root surface. Root hairs are relatively straight, except in *Amaranthus torreyi, Tagetes patula,* etc., wherein they are crooked and irregular shaped. Root hairs are usually unbranched, but rarely branched or forked only in very few plants (Dittmer, 1949). Patriquin *et al.* (1983) noticed branched root hairs in field-grown wheat seedlings, and a greater frequency of these in seedlings inoculated with *Azospirillum brasilense* 245 rather than in uninoculated seedlings. Branching was classified into two types: branches of equal length, the 'tuning-fork deformation', and branches of unequal length. McCully (1987) observed that numerous root hairs in the sheathed region of the grass roots have knobbly curled regions; occasionally some of these hairs were branched at such regions.

b) Dimorphic root hairs

Worrall and Roughley (1976) have reported the presence of two types of root hairs in *Trifolium subterraneum* L. Root hairs growing in the wettest soil (9.5%) are long and slender while the root hairs in driest soil (3.5%) were short and stubby or swollen. The latter kind of hairs were approximately one-third the length and twice the width of those in the former.

In parasitic Scrophulariaceae members, Baird and Riopel (1985) have observed two kinds of root hairs on the same root—the root hairs on haustoria having a thick coating material and the non-haustorial or normal root hairs without any coating material.

2. CYTOLOGY AND ULTRA-STRUCTURE OF ROOT HAIR

The protoplasm within the hemispherical tip appears dense and completely non-vacuolate. This non-vacuolated protoplasm tapers down the sides of the hair from the tip (Belford and Preston, 1961), where there is a concentration of cytoplasmic organelles (ER, ribosomes, mitochondria), especially of many active Golgi bodies. The nucleus is usually located close to the growing tip of the root hair (Sievers, 1963a, b). It was found that in root hairs, protoplasmic streaming is not evident below 0°C; it starts at +6°C and stops at 37°C (Alisa and Soran, 1985).

The cytoplasmic organization at the tip of the growing root hair has been shown to be markedly different from that of the more basal, non-growing regions (Bonnett and Newcomb, 1966). By way of a near median longitudinal section through the apex of a root hair, it was shown that the tip contains only fibrous inclusions, aggregates of ribosomes, and smooth-surfaced vesicles, organelles occupying more basal regions include plastids, mitochondria, dictyosomes, and rough endoplasmic reticulum with occasional dilated cisternae (see **Fig. 2** in Bonnett and Newcomb, 1966). Diagrammatic representation of the ultra-structure of the terminal portion of a root hair has been well depicted by Peterson and Farquhar (1996).

Growing root hairs of *Lepidium sativum* exhibited three zones of differentiation; the tip zone, the zone of vacuolation and the foot zone (Volkmann, 1984). The plasmatic fracture face of the plasma membrane of the tip zone of root hair has shown an irregular fracture plane caused by a large number of blisters coinciding in size and shape with the Golgi vesicles accumulated in the ground cytoplasm near the very tip. The plasma membrane of the vacuolation zone often revealed areas of hexagonally-ordered particles (HOPs). The number of such areas increased with increasing distance from the tip. The plasma membrane of the foot zone showed a constant number of 2000 IMP μm^{-2} and HOPs were never observed in spite of the fact that the cell wall was composed of numerous parallel running cellulose micro-fibrils (Volkmann, 1984). The cytoplasm is very dense and contains many vacuoles, which present autolytic features (Harris, 1979). The cytoplasmic pH of root hairs (in *Sinapsis alba*) is determined to be 7.3 ± 0.2 at neutral external pH (Bertl and Felle, 1985; Felle, 1987).

Cutter and Feldman (1970a and 1970b) have studied the trichoblasts of *Hydrocharis* and stated that during the development of trichoblasts, their nuclei and nucleoli increase in volume. The trichoblasts contain more nucleohistone, total protein, RNA and nuclear DNA than the neighbouring cells. The trichoblasts do not divide and their nuclei become increasingly polyploid due to endomitosis (nearly 4 times). This is the result of the delay in maturation of the developing root hair as it happens with distance from the root tip. According to them, the delay in maturation is probably an essential factor in the differentiation of root hairs. *Hydrocharis* root hairs are noted for their large size, which is probably a reflection of their high degree of polyploidy. But in other plants, root hairs seem to maintain the normal ploidy, as is evident from the studies of Rasheed *et al.* (1990) who found that the plants regenerated from protoplasts of root hairs were cytologically normal.

The root hairs are selected as the experimental cell type to characterize amplification of nuclear DNA because, its physical position in the epidermis correlates with developmental and functional stages and because it is easily harvested for the molecular analysis. The satellite band from root hair DNA represents a highly amplified extra-chromosomal sequences, which may be important in the highly specialized metabolism of the root hairs (Murray *et al.*, 1987). Murray and Christianson (1987) reported a large increase in nuclear DNA content associated with root hairs differentiation. Quantitative microspectrophotometry showed that the root hair nuclei of *Tradescantia* and *Hordeum vulgare* contain substantially more DNA than their respective telophase nuclei, indicating existence of endopolyploidy enhanced by amplification.

Freeze-substitution technique employed to improve the ultra-structural preservation of legume root hairs showed under electron microscopy that there is a 'pyriformis' vesicle in the root hairs, which was previously unreported. Also unique to freeze-substituted material are coated secretory vesicles; smooth plasma membrane profile; mitochondrial ribosomes; long microfilament bundles which are associated with vesicles, mitochondria, coated pits and coated vesicles; microtubule associated filaments; and a pliciform nucleus.

The microtubules in the hairs show net axial orientation of helicoidal pattern, depending upon the species. So, microtubules play some role in microfibril orientation (Traas *et al.*, 1985). The orientation of the cortical microtubule arrays and the last layer of cellulose micro fibrils deposited in the secondary cell wall of *Urtica dioica* root hairs is studied by Amstel *et al.* (1993). It was found that cortical microtubules of individual root hairs showed a preferential orientation, which ranged in the total root hair population from -20 to $+20°$ with respect to the longitudinal cell axis. The cellulose micro fibrils in the secondary wall were organized in 2 step helices. Quantitatively, the majority of the micro-fibrils were oriented in an S helix, while simultaneously, a smaller group was arranged in a Z helix in the same root hair. They concluded that microtubules do not directly control the orientation of nascent cellulose micro-fibrils in this complex texture and the organization of the secondary cell wall texture could be a variant of the organization of the primary cell wall texture.

3. CELL WALL OF ROOT HAIR

The cell wall thickness of root hair is generally less than 1 µm and consists of two layers—an outer layer with random micro-fibrils associated with amorphous wall material largely made up of pectin and hemicelluloses; an inner layer with axially–oriented micro-fibrils without any amorphous material. The micro-fibrils in the outer layer form a loose network. In the outermost tip of the root hairs, the cellulose micro-fibrils are deposited in random fashion, whereas the wall texture becomes more and more complex with increasing distance from the tip (Volkmann, 1984).

The cell wall material is synthesized at the extreme apex of the root hair. The inner layer extends along the cylindrical part of root hair, virtually isolating the outer layer from the protoplast except within the extreme tip portion.

Root hairs in most aquatic plants have a helicoid cell wall texture in the secondary wall (inner layer), whereas in the tips of root hairs, only the primary walls (outer layer) with randomly-oriented micro-fibrils are present. The presence of helicoidal or axially-oriented micro-fibrils in the inner wall layer seems to have some functional significance.

There are an increasing number of helicoidally-arranged lamellae, presenting a counter clockwise mode of rotation that is species-specific in most aquatic plants (Emons and Van Maaren, 1987). In root hairs of *Raphanus sativus*, microtubules formed long-range associations with a constant inter-microtubule space of 50 nm (Seagull and Heath, 1980).

The root hairs of some xerophytes (*Pinus edulis*) have rigid, lignified walls (Coulter *et al.*, 1911). It would be worth investigating the wall structure in case of the dark coloured rigid persistent root hairs of *Anacardium occidentale* and light coloured persistent root hairs of *Ailanthus malabarica*, which would reveal interesting adaptations of these trees.

It is possible to achieve a rapid enzymatic degradation of the cells of the apices of root hairs from a wide range of crop species, thereby exposing the plasma membranes with partial protoplast released whilst maintaining the functional integrity of the plant (Cocking, 1985). Treating of the roots of 24-48 h-old seedlings of *Lotus corniculatus* L. with 1.0% Cellulase YC, and 0.1% Pectolyase Y-23 in 4.2% mannitol solution released protoplasts from the tips of toot hairs within 30-40 seconds of enzyme incubation. Roots from 1000 seedlings yielded 1.7×10^5 protoplasts from which plants were successfully regenerated (Rasheed *et al.*, 1990).

The cell wall of a root hair is very rarely thick walled, as reported in *Gleditsia triacanthos* which turn brown within a few days after being formed (McDougall, 1921). Otherwise, root hairs are most commonly thin walled and slender-like hyphae in fungi.

4. CUTICLE IN ROOT HAIR

Cormack (1937) and Dale (1951) mention the presence of a thin layer of cuticle in the roots of aquatic plants (*Elodea canadensis*). According to Cormack (1937), roots grown in dark (either in water or soil) do not develop any cuticle and produce root hairs plentifully. Whereas, roots exposed to light develop a cuticle and hence, prevent the development of root hairs. The formation of cuticle is said to be associated with chlorophyll formation, photosynthesis and O_2 generation in the light exposed roots. Lee and Priestley (1924) grew a seedling of *Pisum sativum* and placed 4" of its stem underground. Epidermal cells broke through the existing cuticle and formed root hairs, but did not develop any cuticle while underground, thus proving that O_2 produced by photosynthesis in light exposed roots causes saturation of fats and formation of cuticle but does not prevent development of root hair.

On the external surface of the hair, a fatty layer similar in behaviour to cutin and suberin can sometimes be found (Cormack, 1962). It is slightly thicker on full-grown hairs than on younger ones.

Dawes and Bowler (1959) reported the presence of a thin cuticle in the root hair cell wall by using different staining techniques in case of a land plant (*Raphanus*), although they erroneously cited that the presence of cuticle in root hairs is well documented by Cormack (1937) and Dale (1951). In reality, the latter authors only demonstrated the presence of cuticle on the roots but not on root hairs.

Scott (1963, 1966) believed that a cuticle may be present on root hairs in *Vicia*. However, it was not supported by Newcomb and Bonnett (1965) and Bonnett and Newcomb (1966), who conducted research on other plants. No cuticle was visible on the surface of the root hairs in maize and barley.

Studies by Sasson *et al.* (1985) and Emons and Van Maaren (1987) on the cell wall texture of root hairs do not make any mention of the presence of cuticle in the root hair cell wall. Their work seems to be concerned solely with the arrangement of the micro-fibrils in the root hair cell wall in aquatic and terrestrial plants, which has remained a controversial subject. Therefore, whether there is really a cuticle on the root hairs needs to be reinvestigated in different species.

Prominent 'tip callose' is reported in young root hairs of clover, but disappears as the hairs mature (Kumarasinghe and Nutman, 1977). The present author has noticed prominent tip callose even in the mature root hairs of *Syzygium cumini* (**Fig. 1b**).

In parasitic Scrophulariaceae members, the outer wall surface of normal root hairs is smooth and free of extracellular coating material; that of haustorial hairs is coated with an extracellular matrix, which has a granular or papillate appearance (Baird and Riopel, 1985).

Origin and Development of Root Hair

1. ONTOGENY OF ROOT HAIR

Root hairs are simple protrusions from the outer wall of a rhizodermal cell. Typically, each rhizodermal cell can form only one root hair, and only about half of the rhizodermal cells have hairs. However, as an exception in the Pinaceae, root hairs develop as evaginations in the second, or even third layer of cells beneath the root surface, as reported, for example, in *Pseudotsuga menziesii* (Bogar and Smith, 1965 cf. Romberger *et al.*, 1993).

Two distinct types of root hair development have been identified: one in which hairless long cells alternate in the same row with hair-forming short cells (trichoblasts), and the other type with no particular pattern of long and short cells. In the latter type, any cell seems capable of forming a root hair (Reeder and Maltzahn, 1953). This has been found to be a much more reliable taxonomic character, as found in grasses (Row and Reeder, 1957). Row and Reeder studied the root hair development in 82 species of 68 genera of grasses. Alternation of long cells and short cells vs equal-sized ones was found to be a much more reliable character than either root hair position or root hair angle. Where long cells alternate with short cells, the root hairs are near the apical end and grow more or less an acute angle; where the cells are about equal in length, the hairs are most often nearer the centre of the hair-forming cell and perpendicular to the root axis (Row and Reeder, 1957). In some species, all the epidermal cells may be potential trichoblasts; in others, particularly among the grasses, trichoblasts occur in rows and alternate in a regular fashion with non-root hair cells. Trichoblasts are formed as a result of an unequal cell division and the smaller of the two cells formed further developed into root hair (see Figures in Fahn, 1982).

In the lower region of the hair zone, cell differentiation from the protoderm is acropetal. With few exceptions, new hairs do not usually develop among the pre-existing older ones; epidermal cells that have reached a certain degree of development no longer differentiate into root hairs (Cormack, 1962). The new root hairs are constantly formed close to the root tip as the latter continues to grow. Epidermal cells begin to exhibit a protuberance at the apical end of the cell. If the epidermal cell continues to elongate after the appearance of the protuberance, the root hair is found somewhat distant from the apical end of the mature epidermal cell (Fahn, 1982).

In roots cultured in humid air, Jaunin and Hofer (1986) found that the presence of root hairs was not related to root growth. However, maximum hair length and length of the hair zone could be correlated to the elongation rate of the primary root. The emergence of root hairs always took place in the extending zone. In more basal regions, rhizodermis cells could not give rise to root hairs. They also found that cells which are not differentiated into piliferous cells when they cease to elongate cannot subsequently produce hairs.

In *Hydrocharis*, during the development of trichoblasts, their nuclei and nucleoli increase in volume, contain more nucleo-histone, total protein, RNA and nuclear DNA than the neighbouring cells. The trichoblasts do not divide and their nuclei become increasingly polyploid due to endomitosis (nearly 4 times). This process is said to be the result of the delay in maturation of the developing root hair as it happens with distance from the root tip. Root hairs in *Hydrocharis* are noted for their large size—probably a reflection of their high degree of polyploidy (Cutter and Feldman, 1970a, b).

During the formation of root hairs, the induction of tip growth is preceded by a local hydrolysis of the epidermal cell wall (Dazzo et al., 1987, Bakhuizen, 1988). During root hair initiation, a protuberance emerges on the outer tangential wall of the trichoblast or of the hair-forming cell, delimiting an apical and a basal part on the base, which is embedded among the atrichoblasts or the hairless cells. The basal and apical parts do not elongate at the same rate (Cormack, 1962). Root hairs are known to elongate only at the apex where the wall is thinner, softer and more delicate (Troll, 1948; Jackson, 1960; Fahn, 1982). During elongation of the base, the microtubules lie perpendicular to the elongation axis, whereas they are oriented in various directions in non-expanding cell surfaces (Derksen et al., 1986). In the apical region of the hair, the micro-fibrils are oriented at random, enabling isodiametric expansion (Derksen et al., 1986). By freeze-fracture electron microscopy, the presence of 'rosettes', which are believed to contain cellulose-synthase molecules, has been demonstrated in the plasmalemma of elongating root hairs (Volkmann, 1984; Emons, 1985).

The nucleus is located in the protuberance, which relates to its role in controlling local growth in root hairs (Sievers and Schnepf, 1981). Fluorescent microspheres have been used as material markers to investigate the relative rates of cell surface expansion at the growing tips of root hairs in *Medicago truncatula* (Shaw et al., 2000). They proposed three characteristic zones of expansion in growing root hairs. The centre of the apical dome is an area of 1 to 2 µm diameter, with relatively constant curvature and high growth rate. Distal to the apex is a more rapidly expanding region, 1 to 2 µm in width, exhibiting constant surges of off-axis growth. This middle region forms an annulus of maximum growth rate and is visible as an area of accentuated curvature in the tip profile. The remainder of the apical dome is characterized by a strong radial expansion anisotropy, where the meridional rate of expansion falls below the radial expansion rate. According to their findings, the cell cylinder distal to the tip expands slightly over time, but only around the circumference.

Root hairs are used as an excellent system to study the effects of auxin on cell metabolism. They are known to be very sensitive to auxin (Jackson, 1960). The rate of root hairs elongation was significantly increased by 10-13 M indolacetic acid and inhibited by 10-6 to 10-3 M.

2. FACTORS INFLUENCING DEVELOPMENT, DIFFERENTIATION AND DENSITY OF ROOT HAIRS

a) Development of root hairs in aquatic plants

The presence of root hairs on the roots of submerged aquatic plants and the extent of their development have been investigated to some extent. Generally, under submerged condition, hydrophytes fail to develop root hairs due to the heavy cuticulirization of the epidermis. Cormack (1937) showed that in darkness—either in water or soil, or in the presence of light—the roots of *Elodea* have only a thin cuticle and develop abundant hairs. Dale (1951), similarly, found that hairs are produced freely in darkness, or in light in the presence of a high CO_2 tension which prevents the oxidation of the unsaturated fatty acids occurring as a weak film on the epidermis. It is likely that under natural conditions, there is insufficient O_2 in the substrate (or in the root tissues in the absence of photosynthesis) to bring about the formation of a true cuticle, with the result that the epidermal walls remain unhardened and the potential root hairs can develop copiously once the root enters the soil or dark (Cormack, 1949, 1962). Such conditions did not induce root hair production in other species (Sondergaard and Laegaard, 1977).

The supporters of the view that roots principally anchor plants to the substratum present no data on the occurrence of root hairs in aquatic plants (Ludwig, 1881; Noll, 1903; Brown, 1913). According to these authors, aquatic plants, being immersed and all parts being in contact with the water, do not 'need to' increase their absorbing surface by root hairs. Accordingly, they maintained that root hairs are absent or rare in aquatic plants.

Contrary to popular impression, the roots of most submerged hydrophytes (including sparsely rooted plants such as *Elodea*) are reported to develop abundant root hairs (Shannon, 1953), some at least when they penetrate the substrate (Cormack, 1937; Sculthorpe, 1985).

Shannon (1953) studied root hair production in a total of 209 aquatic species, representing 105 genera and 54 families. Most of these species grew naturally rooted in mud, silt or sand on the bottom of shallow water. Many of these also produced roots in water but not in contact with the bottom. Shannon examined for the presence of root hairs on roots growing in water as well as in submerged substrata. In addition, several species of deep lakes, salt water and salt marshes, plants of tidal flats of the Hudson River, plants of acid sphagnum bogs, were also examined for the presence of root hairs. The plants were subjected to the following four conditions and their effect on root hair development was studied:

a) Seeds germinated in water;
b) Plants grown from cuttings in water;
c) Plants grown from cuttings in wet sand; and
d) Plants grown from cuttings in wet sphagnum.

Shannon (1953) reported the existence of root hairs on the roots of many aquatic angiosperms (195 species) either when the plants were kept submerged or when grown in an aerated medium. These investigations did not support the view that root hairs are infrequent on aquatic plants nor that their production is stimulated when the plants are rooted in a substratum. These observations indicated that most aquatic plants, like terrestrial plants, produce abundant root hairs to increase the absorbing surface of their roots.

The aquatic plants that lack root hairs altogether include *Pistia stratiotes* L., *Eichhornia crassipes* (Mart.) Solms, and *Heteranthera zesterifolia* Mart. In these plants, there is some substitution of the hairs by an extensive production of laterals which are initiated very close to the main root apex rather than behind its extension zone (Charlton, 1978). Perhaps, these laterals in *Eichhornia crassipes* were mistaken for root hairs by Bagyaraj *et al.* (1979) who reported the occurrence of VAM fungi in its roots. After this report, the present author also reinvestigated to confirm if root hairs were present in this aquatic plant. The findings showed that there are no root hairs in *Eichhornia crassipes* roots, which confirmed the findings of Charlton (1978). These laterals were actually multicellular in nature and encased in a multicellular single layered capsule (**Fig. 1a**). Further, Bagyaraj *et al.* (1979) included 10 other aquatic species which possessed root hairs. The only exceptional hydrophyte without root hairs was *Cyperus eleusinoides*. Later, while studying VAM in the submerged aquatic plants of New Zealand, Clayton and Bagyaraj (1984) reported that out of 29 species studied for root hairs and VAM, 15 species had abundant root hairs, both among the deep water and shallow water plants, while 14 species either had sparse or no root hairs. Clayton, in his personal communication to the present author, wrote that he had not studied the root hair aspect and he was unable to send any photographic evidences on the presence of root hairs in these aquatic plants.

It is possible that the presence of root hairs was usually recorded with the help of a dissecting microscope (as for example St. John, 1980) and with which it is not possible to make out whether the hairs are really single celled or multicellular laterals, as in case of *Eichhornia*. Hence, the presence of root hairs in submerged parts of aquatic plants or hydrophytes was doubtful. In order to confirm the presence of root hairs in aquatic plants, the present author conducted further studies in partly-emergent hydrophytes, namely *Ludwigia* and *Alternanthera sessilis*.

Ludwigia spp. (= *Jussiaea*) usually produce two types of root. Downward-growing roots are called 'mud roots' which are long and bare of root hairs. On the other hand, the upward-growing roots called 'aerial roots', which in some species develop into spongy bulbous floats of a aerenchymatous structures (as in *Ludwigia adscendens* or *Jussiau repens*) that lack root hairs. In *Ludwigia octovalis*, the present author has observed that the aerial roots, instead of developing into floats, grow above the water level and develop whitish dense root hairs.

Alternanthera sessilis, which is a semi-aquatic or an emergent aquatic plant and often straggles over aquatic weeds such as *Eichhornia crassipes*, exhibits two kinds of roots. Upper nodal roots, which are produced just below the water level, exhibit dense and long root hairs, while the roots produced at lower nodes submerged in deeper water depths (0.5—1.0 m) are without any root hairs (Bhaskar and Kushalappa, unpublished data). The roots produced from nodes above the water level also developed root hairs but much shorter and fewer than the root hairs produced just below water level. As these roots became submerged and further branched out, root hairs gradually obliterated. This can be attributed to the linear gradient that exists with regard to dissolved O_2 content in the water from surface to the bottom of the tank.

The above finding shows that in aquatic plants, root hairs are mostly produced in the roots produced above or close to water surface (as in *Ludwigia octovalis*) or in water closer to air with more dissolved O_2 and that O_2 is essential for the development of root hairs. This is contrary to the observation of Dale (1951), who presumed that O_2 presence would oxidize fatty acids to produce cuticle, which prevents root hair formation. While in *Ludwigia*, although roots are exposed to air (or directly to gaseous O_2) or closer to air as in *Alternanthera sessilis*, it does not prevent root hair development. Further, Bhaskar and Kushalappa (unpublished data) have found that the quality of water, particularly the cleanness and freshness of water, significantly affected the development of root hairs in *Alternanthera sessilis*. When plants were grown in polluted stagnant water with high BOD (Biochemical Oxygen Dimond) and clean fresh water with low BOD and were then studied, it was found that plants grown in the latter case only developed very long and normal root hairs, whilst those in former did not develop any. All these findings indicate that root hairs are more responsive to O_2-rich environment.

b) Role of O_2 in root hair development in terrestrial plants

The role of O_2 in root hair formation in terrestrial plants is also evident from the fact that most roots generally produce root hairs much more abundantly in moist air than when submerged in water. Some species are completely unable to form hairs in water (von Guttenberg, 1968). Root hairs often most conspicuously develop on roots growing in soil voids with air gaps of appreciable size (Russel, 1977). Further, more dense root hairs have been found to develop in relatively dry and porous soils than in too wet soils (Pfeffer, 1897 cf. Whitaker, 1923; Bhat and Nye, 1973; Mackay and Barber, 1985, 1987). In tissue cultures, the present author has observed that only those roots that grew outside the culture medium possessed dense root hairs rather than on those roots which were submerged in the medium (**Fig. 11**).

As observed by the present author, viviparously-germinating seeds, as in fresh pea (*Pisum sativum*), are seen to have dense root hairs on the radicle growing into the air spaces inside the pod (**Figs. 6a and b**). In a tomato plot which was excessively irrigated, plants produced new roots outside the soil surface which possessed dense root hairs while the roots inside the soil were devoid of it (**Fig. 9**).

The oxygen concentration of nutrient solution has been shown to promote root hair development in tomato (Yang *et al.*, 1991). It was found that the number of root hairs, length of root hairs and the length of root hair zone in the 2 cm region above the root tip was greatest at 10 or 20 p.p.m. in tomato. Root hair growth above this region was better at 20 or 40 p.p.m. oxygen concentration.

Thus, from the above evidence, it can be seen that hydrophytes and terrestrial plants exhibit a distinct adaptive difference with regard to the requirement of either oxygenated water medium (in case of hydrophytes) or air (in case of terrestrial plants) to develop root hairs, but in both, oxygen as an essential requirement for development of root hairs is evident.

Thus, root hairs can be classified into the following types:

(1) Aerial root hairs; and
(2) Aquatic root hairs.

Aerial root hairs are produced in the presence of air and are sensitive to direct contact with water or any liquid medium (e.g. majority of mesophytic land plants and rarely in aquatic plants). Among the aerial type of root hairs, root hairs may be of two kinds—one where they are ephemeral or become flaccid if exposed to dry air with less moisture and the other kind of root hairs which are persistent and remain rigid. Cell wall structure may be responsible for such a difference.

Aquatic root hairs are produced either in oxygen-rich water or in surface water closer to air or oxygen source and are sensitive to direct exposure to air (e.g. hydrophytes). It will be interesting to study and compare the cell wall structure in these different types of root hairs and find out their structural and functional adaptations.

c) Effect of ambient environment on root hair development

Mer (1879) stated that increased humidity favours root growth but inhibits root hair formation and that root hairs are not commonly formed in aqueous media. Farr (1928b) showed that except in solutions of approximately optimum molar concentration, there is a temporary cessation of growth of root hairs after immersion. Farr (1928c) also showed that root hairs which have begun to develop in air, when placed in even slightly toxic calcium chloride solution, became modified in form by swelling, inflation, branching or curvature. In very toxic solution, they burst and collapsed.

According to Uphof (1962), an optimal level of humidity seems to be one of the essential factors for the development of root hair zone, but Vartanian *et al.* (1983) are of the opinion that humidity seems to effect the elongation of the hairs but not the differentiation of the protoderm into hair cells. Most roots generally produce hairs much more abundantly in moist air rather than when submerged in water. Some species are completely unable to form hairs in water (von Guttenberg, 1968). The role of oxygen in root hair formation has been reported for various species (Cormack, 1962). Root hairs may be induced in an aqueous medium by increasing its O_2 content (Hofer, 1991). Sun (1955), in his experiments with soybean, found that root hairs are sparsely developed on aerated primary roots. Roots grown in a non-aerated solution produced an abundance of root hairs. However, the present author has observed that when the just-germinated soybean seeds were subjected to water submersion and aerated treatments, radicles exposed to air developed dense root hairs but failed to do so under water submersion (**Fig. 8d**).

Rosene (1954), during the course of her physiologic studies with onion roots, found that the adventitious roots of this plant do not readily produce root hairs when cultured on aerated Hoagland nutrient solution nor when grown experimentally in moist air, but produce hairs abundantly when they come in contact with filter paper moistened with a culture solution.

Root hairs of the tree seedlings grown in vermiculite seemed to be longer compared to seedlings grown in paddy soil and granite soil (Poerwanto *et al.*, 1987).

Teak seedlings grown in clayey soils were devoid of root hairs but developed well when transferred to a sand medium (Bhaskar, unpublished data). The rhizotron study has shown no differences in the percentage of roots having root hairs regardless of water stress treatments imposed 20 to 50 and 50 to 80 days after planting (DAP). Thirty-seven to 43% of the roots had root hairs. In the field, peanut root systems growing within dialysis root bags in a lifton loamy sand (fine-loamy, siliceous, thermic Plinthic Kandiudults) showed a significant increase in root hair formation from 7% at 70 DAP to 31% at harvest. In the greenhouse, soil type was found to affect the percentage of roots having root hairs. Roots growing in the Tifton, sand:fritted clay mixture, and Greenville soils had 47, 17, and 14% root hairs, respectively (Meisner and Karnok, 1991).

d) Role of mineral nutrients in root hair development

Soil moisture and soil phosphorus levels seem to affect the length and density of root hairs (Bhat and Nye, 1973; Mackay and Barber, 1984). Of these two factors, soil moisture is the most dominant one (Mackay and Barber, 1985). As soil dries, the root hair growth of maize roots has been found to increase, even at very high soil P levels (see **Fig. 3** in Mackay and Barber, 1985). Under dry soil conditions (M_0), the percentage of total root length with root hairs and the density of root hairs were 1.4 to 1.8 and 2.0 to 4.2 times greater, respectively, than under wet conditions (M_1) on three types of soils. Drying soils from M_1 to M_0 after 7 days resulted in an increase in root hair initiation and growth on new roots.

A greater amount of Ca^{2+} was found in the hair cells rather than in the hairless cells (Jaunin, 1988). Calcium is known to be an essential key constituent of culture media for root hair formation (Ekdahl, 1953; Cormack, 1962; Tanaka and Woods, 1972; Ewens and Leigh, 1985; Jaunin and Hofer, 1988). The role of extracellular Ca^{2+} in root hair tip growth has been reported in *Arabidopsis thaliana* (L.) Heynh (Schiefelbein *et al.*, 1992). Root hair length was found to be dependent on the concentration of Ca^{2+} in the growth medium, with maximum length achieved at Ca^{2+} of 0.3-3.0 mM. The direction of the gradient indicated a net influx of Ca^{2+} into root hair cells. No gradient was detected either near the sides of the root hairs or at the tips of non-growing root hairs. These results indicated that Ca^{2+} influx through plasma membrane Ca^{2+} channel is required for normal root hair tip growth. Although the elongation rate of *Triticum* root hairs seemed to be independent of the Ca^{2+} concentration in the culture medium (Ekdahl, 1953), a tip-to-base gradient of this ion was found in the tube of the hairs of several species (Cormack, 1962; Reiss and Herth, 1979). In *Arabidopsis thaliana*, emerging root hairs showed an elevated (Ca^{2+})c level (> 1 µM) in their apical cytoplasm, compared with 100-200 nM in the rest of the cell. Rapidly-elongating root hairs exhibited a highly localized, elevated (Ca^{2+})c at the tip, although non-growing root hairs did not exhibit this gradient (Wymer *et al.*, 1997).

However, stimulation of the production of root hairs at low nutrient availabilities has been reported frequently (Baylis, 1970, 1972; Bhat and Nye, 1974; Mackay and Barber, 1984, 1985; Ewens and Leigh, 1985), although the precise regulation of their growth by environmental and internal factors remains obscure. Root hair growth in some grasses was the most responsive to low N availability (Robinson and Rorison, 1987).

e) Effect of pH on root hairs

Even the pH of the medium has been found to affect root hairs. Lupins and peas were grown hydroponically in solutions buffered at pH 5.2 or 7.5. In lupin roots, pH 7.5 caused disintegration of the root surface and impaired root hair formation (Tang *et al.,* 1993). However, in halophytes, the number of root hairs is reported to increase with the salt concentration in the soil, up to an optimal concentration (Uphof, 1962). Increasing salinity, leading to a reduction in the number of root hairs and the formation of a mucilaginous layer around the roots has been reported in lucerne (*Medicago sativa* L.) (Lakshmi-kumari *et al.,* 1974). Even the sparse numbers of root hairs present appeared to be short, stubby and bulbous. Carbonate and bicarbonate even at 0.2 per cent were shown to inhibit the formation of root hair development. It appears that the formation of root hairs was not affected up to pH 8.0, beyond which a reduction in root hair formation set in.

f) Role of microbes on root hair development

Certain microbes like *Azospirillum* are reported to influence root hair development. As per research, *Azospirillum* stimulates the proliferation of root hairs (Kapulnik *et al.,* 1985; Bothe *et al.,* 1992). Incubation of germinated wheat cv. Ralle seeds in a hydroponics system with *Azospirillum* slightly increased the formation of root hairs and it was suggested that the growth promotion effect was due to nitrite formation (Bothe *et al.,* 1992). Broek *et al.* (1993) observed that in wheat, *Azospirillum brasilense* was mainly found in the root hair zones during the first days of association and their further proliferation was dependent on the nitrogen status of the nutrient solution.

Inoculation of pea plants with *Penicillium bilaii* in growth pouches is reported to have resulted in a 22% increase in the proportion of root containing root hairs and a 33% increase in the mean root hair length (Gulden and Vessey, 2000).

g) Role of hormones in root hair development

The massive production of root hairs in a number of species is another interesting response of roots to ethylene. It is not yet known whether ethylene is involved in the natural development of root hairs (Reid, 1987). In a hydroponics system, when incubated with IAA, it is reported to enhance the root hair formation in germinated wheat cv. Ralle (Bothe *et al.*, 1992).

Root hair development has been intensively studied in *Arabidopsis thaliana* (Pitts *et al.*, 1998; Cao-XiaoFeng *et al.*, 1999). Pitts *et al.* (1998) studied the root hair phenotype of a number of auxin- and ethylene-related *Arabidopsis* mutants in order to investigate the role of auxin and ethylene in root hair development. They showed that mutants deficient in either auxin or ethylene response had a pronounced effect on the root hair length. Treatment of wild-type, axr1 and etr1 seedlings with the synthetic auxin 2,4-D or the ethylene precursor ACC, led to the development of longer root hairs, compared with untreated seedlings. They concluded that both auxin and ethylene are required for normal root hair elongation. Root hairs were found to develop on trichoblasts located over the anticlinal (radial) walls of the underlying cortical cells, while non-hair cells developed on atrichoblasts overlying the periclinal (tangential) walls of cortical cells. Dark-grown wild-type seedlings, which produced little ethylene, were found to be largely lacking in root hairs, while the exogenous treatment of dark-grown plants with either ethylene or ACC restored the development of root hairs in cells overlying the anticlinal cortical cell walls, indicating that cells in this position are more sensitive to ethylene than atrichoblasts (Cao-XiaoFeng *et al.*, 1999).

3. GENETIC CONTROL OF ROOT HAIR DEVELOPMENT

Genetic control of root hair development has been studied in *Arabidopsis thaliana* (Shiefelbein and Somerville, 1990). Visual examination of roots from 12,000 mutagenized seedlings led to the identification of more than 40 mutants impaired in root hair morphogenesis. Mutants from four phenotypic classes have been characterized in detail, and genetic tests show that these result from single nuclear recessive mutations in four different genes designated RHD1, RHD2, RHD3, and RHD4. The RHD1 gene product appears to be necessary for proper initiation of root hairs, whereas the RHD2, RHD3, and RHD4 gene products are required for normal hair elongation.

A gene affecting tomato root hair production in a solution culture was reported by Hochmuth (1986). Nambiar *et al.* (1983) reported the complete absence of root hairs in some 24 groundnut mutant lines. Root hair cells could be obtained more easily from cultures derived from *Glycine max* var. Acme than from var. Harosoy (Hermina and Reporter, 1977).

The various characteristics of root hairs such as their presence or absence, number, length, etc., vary in different species, varieties and even cultivars, which is determined by genes. For example, cultivar Salka in Barley having more lengthier root hairs absorbed twice more P from rhizosphere soil than cv. Zita with shorter root hairs (Gahoonia and Nielsen, 1998).

4. BIOTECHNOLOGY WITH ROOT HAIR

It is possible to achieve rapid enzymatic degradation of the cell walls of the apices of root hairs from a wide range of crop species. Thereby, it is possible to expose plasma membranes with partial protoplast release whilst maintaining the functional integrity of the plant. This exposure of plasma membranes provides a port of entry into the plant for a wide range of investigations, including genetic manipulations (Cocking, 1985).

Rasheed *et al.* (1990) have successfully regenerated plants from root hair protoplasts of *Lotus corniculatus* L. Treatment of the roots of 24-48 h-old seedlings, when treated with 1.0% Cellulase YC, and 0.1% Pectolyase Y-23 in 4.2% mannitol solution, released protoplasts from the tips of root hairs within 30-40 seconds of enzyme incubation. Ten per cent of the protoplasts divided in order to form cell colonies when cultured at 1.0×10^5/ml in droplets of KM8P medium with 0.6% Sea Plaque agarose. Colonies formed callus on UM agar medium and protoplast-derived tissues produced shoots on B5 medium containing 0.05 mg benzyladenine/litre.

Root hairs can be induced through microbial inoculations to the growing medium. Hairy roots with numerous root hairs were induced by inoculating *Agrobacterium rhizogenes* to leaf discs of horse radish (*Armoracia lapathifolia*) and cultured on a solid medium (Noda et al., 1987). The excised hairy roots grew vigorously in the dark and exhibited extensive lateral branches and when moved into the light, numerous adventitious buds thrust out of the roots within about 10 days, and developed into complete plants. Inoculation of tomato seedlings (*Lycopersicum esculentum*) with a different *Azospirillum brasilense* inoculum had a marked effect on the proliferation of root hairs at a specific root zone (1 cm from root tip) (Okon, 1988).

Large number of root hairs have been induced in suspension cultures from soybean roots (Hermina and Reporter, 1977). Root hair proliferation was obtained when cells were placed in medium which was lacking 2,4-D and kinetin.

Physiological Functions of Root Hairs: Earlier Theories

1. ROLE OF ROOT HAIR IN WATER AND MINERAL UPTAKE

Right from the beginning, two main functions have been attributed to root hairs: a role in water and mineral nutrient uptake, and in adhesion properties between the root and its surroundings. Root hairs are generally considered to increase the surface area of roots and thereby, increase absorption of water and mineral nutrients from the soil (Dittmer, 1937; Rosene, 1943; Cailloux, 1943; Bowen and Rovira, 1969; Reynolds, 1975; Lamont, 1982; Mackay and Barber, 1985, 1987). According to Hofer (1991), at least for the emerged plants, the possibilities of water and ionic exchanges with the environment depend in part on the number of hairs and on their surface area. A finely divided root system with abundant root hairs is a 'simple' but efficient device to increase the absorptive area. A number of plants have root clusters or bunches of hairy rootlets, which proliferate in the decomposing litter, and between the raw litter and humus horizon, which is a store of water and nutrients and the site of most nutrient release (Lamont, 1982). The coleorhiza hairs in grasses, which are similar to root hairs, are also believed to play a role in increasing the surface area for water absorption (Debaene-Gill et al., 1994).

a) Water absorption

Ohlert (1837) was the first researcher to prove that roots, mainly the 'whitish zone' above the yellowish one, absorb water. Then, Meyen (1838) proposed that roots absorb water mainly by their root hairs. For the first time, Cailloux (1943) showed by actual measurements on individual root hairs, that they possess the ability to absorb water. Rosene and Walthall (1949, 1954) showed that absorption of water is mainly performed by young root hairs and slows down as the root hairs mature. Water uptake efficiency was correlated with root hair length (Caradus, 1979; Itoh and Barber, 1983).

Further, it was shown that water uptake in a root hair was confined largely to the tip, i.e. the region that contains a non-vacuolate mass of dense protoplasm. No absorption was detected in the lower regions of the root hair facing the central vacuole (Cailloux, 1953, 1972). Water absorption by *Avena* root hairs was not proportional to the hair surface in contact with water; it occurred only in a small region where the main cytoplasmic mass is located, not in the region facing the vacuole (Cailloux, 1974). When 127.5 μm of the tip of root hair is immersed in a capillary, it absorbed, whereas the same root hair immersed at a depth of 255 μm excreted (Cailloux, 1950).

Newman (1974) postulated that root hairs enhance water uptake through their ability to enter smaller sizes of water-filled pores in the soil that the whole root could not directly harvest. According to Tinker (1976), root hairs are impor7tant in water uptake because they help in maintaining liquid continuity across the soil-root interface.

Natural electric fields were detected and measured in the medium near root hairs of barley seedlings with the aid of an extra-cellular vibrating electrode (Weisensell *et al.*, 1979). Root hairs as well as the entire root drew large steady currents through themselves. Current was found to consistently enter both the main elongation zone of the root as well as the growing tips of elongating root hairs. It also indicated that much of the current consists of H^+ ions. Altogether, H^+ ions seem to leak into growing cells and to be pumped to the non-growing ones. Their result indicated a relationship between currents and growth; the currents create and maintain gradients of charged cytoplasmic and membrane components which support the localized growth of cells and tissues.

It is a well-established fact that most plants occupying wetland or aquatic habitats (hydrophytes) lack root hairs (Sculthorpe, 1985; Curran *et al.*, 1986). The generalized absence of root hairs in hydrophytes is ordinarily attributed to the submergence of part or the entire plant body in water. Thus, many physiologists felt that hydrophytes did not evolve any special organs to facilitate water absorption.

According to Mackay and Barber (1987), the increase in root hair growth in dry soils rather than in wet soils may be an attempt by the plant to maintain liquid continuity around the growing root and to provide greater root surface for nutrient absorption as the rate of nutrient diffusion to the roots decreases in the drier soil.

b) Ion exchange or mineral uptake

Auto-radiographic evidence suggests that nutrients are also taken up by root hairs (Lewis and Quirk, 1967; Bhat and Nye, 1973; Fohse *et al.*, 1991). Phosphorus (P) uptake is shown to vary in different species, according to root hair length and density. Species having fewer and smaller root hairs (as in carrot, onion, *Coprosma robusta*) do not have a high P uptake efficiency per unit area of root (Sumio and Barber, 1983). In such cases, plants often develop larger root areas or form mycorrhizal associations that serve to harvest the required P (Sumio and Barber, 1983; Baylis, 1970). Barley and Rovira (1970) found a significantly greater uptake of phosphorus in the presence of root hairs rather than their absence. The root hair density for wheat was low compared with Russian thistle, tomato, and lettuce. Consequently, wheat roots took 2.5 days to deplete one-half the P in the root hair zone, whereas lettuce needed only 0.6 day. Rovira and Bowen (1968) found that P uptake occurred rather uniformly over a 7 cm portion of wheat roots, while chloride and sulfate absorption was localized closer to the root apical meristem. They also found wheat roots to have two major sites for absorption, one 4-7 mm behind the meristem and another 20-30 mm behind the apex. Bole (1973) found no significant relation between the number of root hairs and the total phosphorus uptake by different wheat cultivars, but Itoh and Barber (1983) discovered a very close relationship in the difference between predicted and observed phosphorus uptake and root hair morphology in a study using six plant species and showed that the difference could be explained by P uptake by root hairs. A mechanistic model was developed and verified with an experiment using 6 species that varied widely in root hair length, density and radius, which illustrated that root hairs contribute significantly to phosphorus uptake. Length of root hairs, root hair density and root hair radius all influenced predicted phosphorus uptake, with root hair length being particularly significant. In both laboratory and field experiments, barley cv. Salka, with longer root hairs (average 1.10 mm), absorbed twice more P from rhizosphere soil than cv. Zita, with shorter root hairs (average 0.63 mm). Results suggest that P uptake from low-P soil can be enhanced by selection of crop cultivars for longer root hairs (Gahoonia and Nielsen, 1998).

Pressure probe tests have shown that the hydraulic resistance of root hairs is similar to that of hairless epidermal cells and cortical cells (Jones *et al.*, 1983). Studies on the function of root hairs and adjoining hairless epidermal cells in the higher aquatic plant (*Trianea bogotensis*) have shown that net K^+ influx was virtually restricted to the root hairs.

The function of root hairs and adjoining hairless epidermal cells in the higher aquatic plant *Trianea bogotensis* seems to be determined by the pattern of plasmodesmata, connecting each epidermal cell to the underlying cortex. Net K^+ influx was observed to be virtually restricted to the root hair cells and related to the frequency of plasmodesmata in the inner tangential wall, which was some 20-fold greater than in the ineffective hairless epidermal cell (Drew, 1987).

Polarization of transport systems and the functional specificity of root hairs was studied by Vakhmistrov and Zlotnikova (1990), who found that cationic dyes were absorbed by growing root hair tips, whereas anionic dyes were taken up by root hair bases.

Even today, physiologists feel that after years of speculation, nutrient transport by root hairs has been clearly demonstrated at the physiological and molecular level, with evidence for root hairs being intense sites of H^+-ATPase activity and involved in the uptake of Ca^{2+}, K^+, NH_4^+, NO_3^-, Mn_{2+}, Zn_{2+}, Cl^- and $H_2PO_4^-$ (Gilroy and Jones, 2000).

2. ROLE OF ROOT HAIR IN ROOT NODULE FORMATION

The role of root hairs in the entry of nitrogen-fixing microorganisms into the root systems, as in the case of N_2-fixing leguminous plants, has been very well established. Although exceptions exist, the importance of root hairs for nodulation in most leguminous species is evident from the complete absence of root nodules in groundnut mutant lines, which also lack root hairs (Nambiar et al., 1983). None of the 24 stabilized non-nodulating lines produced any root hairs. In *Arachis hypogea*, Rhizobia were observed to enter the root at the junction of the root hair and the epidermal and cortical cells (Chandler, 1978) as though the root hairs are not involved in nodule formation. However, Nambiar et al. (1983) concluded that nodulation failure in some of the genetic lines of groundnut is associated with the absence of root hairs.

There are no root nodules in *Gleditsia triacantha* of the Leguminosae as root hairs become thick walled very quickly (McDougall, 1921).

The morphology of root hair also seems to affect infection and nodulation. In *Trifolium subterraneum* L., only the long and slender root hairs produced under the wettest soil moisture were infected and developed into nodules, whereas the short and swollen root hairs growing in the driest soil were not infected at all. However, the short and stubby hairs which resumed normal growth on rewatering were infected (Worrall and Roughley, 1976).

It is also reported that *Azospirillum*, which is known to stimulate proliferation of root hairs (Kapulnik *et al.*, 1985; Bothe *et al.*, 1992), also increases the probabilities of infection by *Rhizobium* and hence an increased number of root hairs increases the chances of nodule formation (Eli Yahalom *et al.*, 1987).

a) Site of infection on root hair

According to Bauer (1981), the Rhizobium infection of legumes starts at the site where a root hair is about to be formed. Wood and Newcomb (1989) studied the infection of alfalfa (*Medicago sativa*) root hairs by *Rhizobium meliloti* and found that uninoculated root hairs grew and matured over a 10-h growth period. The nucleus migrated from a position opposite that of root hair protrusion at the initiation to the base of the root hair protrusion, then into the growing root hair. If root hairs were inoculated during the first 2 h of growth after initiation, root hairs deformed into a tight curl as the tip developed and later demonstrated typical infection thread formation. Root hairs older than 6 h at the time of inoculation formed a variety of growth deformation, including ballooning and elongate, spatulate, spiraling or intertwined growth. Infections in this population of root hairs were rare. Therefore, root hair deformation is induced in a small zone of the root containing root hairs that have almost stopped growing. This suggests that only at a specific stage of development are root hairs able to deform. Lotocka *et al.* (2000) have shown that infection occurred via the curled root hair. Rhizobia penetrated the cell wall, and then multiplied inside the root hair. Some rhizobia escaped the penetration site before the host cell built a new wall around it. Escaped bacteria passed to the base cell of the root hair, where cell wall penetration and matrix excape occurred again.

b) 'Nod' factor and root interaction

The Rhizobium-legume interaction starts with the exchange of signal molecules between both partners. Flavonoides secreted by the roots of the host plant trigger the expression of the nodulation (nod) genes of Rhizobium, resulting in the synthesis of specific lipo-oligo saccharides named Nod factors (Fisher and Long, 1992; Spaink, 1992; Denarie and Cullimore, 1993). Nod factors play a key role in the induction of early steps of nodulation (see **Fig. 2** in Heidstra *et al.*, 1994 and **Figs. 3, 6 and 7** in Ardourel *et al.*, 1994 and Plate 1 in Worrall and Roughley, 1976).

According to Heidstra *et al.* (1994), who monitored root hair deformation with a video camera, 5 to 10 minutes of Nod factor-root interaction appears to be sufficient to induce root hair deformation. An increased cytoplasmic streaming occurs within 30 minutes. The first deformation is visible within 1 h., and after 3 h about 80% of the root hairs in a small susceptible zone of the root become deformed. This zone encompasses root hairs that have almost reached their maximal size. The Nod factor accumulates preferentially to epidermal cells of the young part of the root, but is not restricted to the susceptible zone. After about 1 h, the tips of the root hairs start to swell, which was more pronounced after 1.5 h. After 2 h, the polar tip growth was initiated from the swollen root hair tips and after 3 h, the root hairs develop a typical deformed appearance. Thus, Nod-factor-induced root hair deformation is a rapid response of the plant. It is possible that Nod factors induce tip swelling by targeting hydrolytic enzymes to the tips of root hairs. The tip swelling is followed by new polar growth from the swollen root hair tip.

c) Histochemistry of root hair infection in legumes

There seems to be a different histochemistry of legume root hairs that are the target cells of Rhizobia. According to Ardourel *et al.* (1994), *Rhizobium meliloti* produces lipochito-oligosaccharide nodulation Nod Rm factors required for nodulation of legume hosts. Nod Rm factor structural requirement seems to be clearly more stringent for bacterial entry than for the elicitation of developmental plant responses.

The interaction between Rhizobium lipopolysaccharide (LPS) and legume roots (*Trifolium repens*) has been examined by Dazzo *et al.*

(1991). They showed that purified LPS from R. *trifolii* 0403 bound rapidly to root hair tips and infiltrated across the root hair wall. Infection thread formation in root hairs was promoted by preinoculation treatment of roots with R. *trifolii* LPS at a low dose. Most infection threads developed successfully in root hairs pretreated with R. *trifolii* LPS, whereas many infections aborted near their origins and accumulated brown deposits if pretreated with LPS from R. *meliloti* 102F28. LPS from R. *leguminosarum* 300 also caused most infection threads to abort. Other specific responses of root hairs to infection-stimulating LPS from R. *trifolii* included acceleration of cytoplasmic streaming and production of novel proteins.

Specific soluble proteins have been found in root hairs with molecular weights of 13, 21, 34, 38 and 42 K^{Da}. Additionally, proteins with molecular weight of 12, 20, 69 and 74 K^{Da} were significantly enriched in root hairs as compared to those roots without root hairs. The special proteins present in legume root hairs may enable recognition by Rhizobium; otherwise Rhizobium could have formed nodules on non-leguminous trees also. Mort and Grover Jr (1988) reported that root hairs of legumes have very similar compositions, whereas those of different families varied widely.

Molecular mechanisms by which Rhizobium bacteria adhere to plant root hairs have been reviewed by Smit *et al.* (1992). According to them, rhizobial attachment to plant root hairs appears to be a two-step process. A bacterial Ca^{2+}-binding protein, designated as rhicadhesin, is involved in direct attachment of bacteria to the surface of the root hair cell. A second step results in accumulation and anchoring of the bacteria to the surface of the root hair.

It is demonstrated by Ohlendorf (1985) that the bacteria directly enter the root hairs, which is in disagreement with the 'invagination' hypothesis of Nutman (1956). According to Gerrit *et al.* (1986), Rhizobia attach to the tips of root hairs. Clumps of bacteria, designated as caps, are formed at the tips of root hairs. A minimal number of attached bacteria is needed before a cap can be observed (in about 90 minutes). This cap formation is inhibited in acid soil conditions (Munns, 1968). Cocking (1985) has shown that the cell walls of the apices of root hairs can be easily degraded enzymatically to expose plasma membrane, hence the root hair apices form the main port of entry for the microorganisms. Lectins have been reported to be present at the surface of root hairs, which are believed to promote the aggregation of rhizobia in the infection zone (Brewin and Kardailsky, 1997).

For all plant species studied, the presence of *Rhizobium* stimulated the synthesis of callose in the root hair. Electron micrographs of root hairs showed regions of presumptive callose wall thickening, sometimes associated with vesicles that may have been concerned with transporting material to the site of wall synthesis. All hairs to which bacteria were attached bore peg-like projections on their outer surfaces, which may be protrusions of pectic material or callose.

Sieberer and Emons (2000) studied the cytoarchitecture and the pattern of cytoplasmic streaming changes during the development of root hairs of *Medicago truncatula* and after a challenge with nodulation (Nod) factors. They observed that the organelle movement in the swelling tip increases up to the level normal for circulation streaming, and the number of strands with moving organelles increases.

d) Root hair infection in actinorrhizal plants

The situation in actinorrhizal plants is more complex because some species are obligatorily infected by *Frankia* through the root hairs, whereas others may be infected through the epidermal cell walls and still others by a combination of root hairs and epidermal cell walls (Baker, personal communication). Miller and Baker (1986) showed that two pure cultured *Frankia* strains could infect *Elaeagnus* by an intercellular penetration and *Myrica* by the root hair infection and neither *Elaeagnus* nor *Myrica* could be infected by both the mechanisms. According to Callaham *et al*. (1979), Berry and Torrey (1983), Callaham and Torrey (1977), the roots of actinorrhizal plants uninoculated with *Frankia* do not develop deformed root hairs, which is crucial as only deformed root hairs become infected. Huss-Danell (1997) showed that *Frankia* infects the roots via root hairs in some hosts or via intercellular penetration in others.

Berry *et al*. (1986) have made a detailed study on the fine structure of root hair infection, leading to nodulation in the *Frankia-Alnus* symbiosis. The *Frankia* hyphae were found to extend through the root hair wall in a highly deformed region of the hair. The primary wall fibrils of the root hair appeared disorganized at the site of penetration. There was extensive secondary wall formation in the infected hair. At the site of penetration, root hair cell wall ingrowths occurred that were structurally consistent with transfer cell wall formation. The ingrowths were continuous with the encapsulating wall layer surrounding the *Frankia* hypha. The host cytoplasm was rich in ribosomes, secretory products, and organelles, including Golgi bodies, mitochondria, plastids, and profiles of endoplasmic reticulum. The root hair cytoplasm was senescent, however, and a callosic plug appeared to surround the pathway of infection. According to these authors, the changes in the host root hair associated with infection by *Frankia* appear to be a function of cell specialization indicating that the root hair is metabolically active and differentiated, perhaps for a secretory or transport function. According to the findings of Ghelue *et al.* (1995), many actinorhizal plants are infected via deformed root hairs. Factor(s) eliciting root hair deformation in actinorrhizal symbioses have been found to be released into the culture medium. No correlation was found between *Frankia* strains belonging to different host specificity groups and their ability to form root hairs on *Alnus glutinosa*. However, strains not able to deform root hairs were also unable to nodulate.

Structural and functional comparison of Frankia root hair deforming factor and rhizobia Nod factor was made by Ceremonie *et al.* (1999), who tested biochemical and functional analogies using Frankia. According to them, rhizobia Nod factors and Frankia root hair deforming factor are two structurally divergent symbiotic factors.

Massive root hair deformation of nodulating and non-nodulating plants is reported to be caused by certain noninvasive bacteria, including *Pseudomonas* spp. (in case of actinorrhizal plants) and *Azospirillum* spp. (in cereal crop plants). The strain-specific effects of *Azospirillum* on root hair deformation in wheat has been reported (Patriquain *et al.*, 1983). Root hair deformation effects were qualitative and quantitative in nature, which is true also of strain-specific effects of *Rhizobium* on root hair deformation in legumes. These bacteria appear to facilitate infection and nodulation of host plants.

The root hair infection and number of nodules are also greatly influenced by the pH of the soil. At higher pH levels, the number of infected root hairs and number of nodules were reduced in lucerne (*Medicago sativa* L.), as reported by Lakshmi-kumari *et al.* (1974). Under increasing salinity, even the sparse numbers of root hairs present did not show curling, characteristic of early phases of nodule formation.

3. ROOT HAIRS AND MYCORRHIZA

Mycorrhizae occur in 83% of dicotyledonous and 79% of monocotyledonous plants (Trappe, 1987). All Gymnosperms are reported to be mycorrhizal (Newman and Reddell, 1987). Ectomycorrhizae are reported among 25 families of vascular plants, for example, Dipterocarpaceae (98%), Pinaceae (95%), Fagaceae (94%), Myrtaceae (90%), Salicaceae (83%), Betulaceae (70%). Vascular-arbuscular mycorrhizae (VAM) are more prevalent than the ectomycorrhizae.

a) Root hairs: VAM relationship in terrestrial plants

Baylis (1975) suggested that primitive angiosperm roots, typified by those of the order Magnoliales, are especially dependent on vesicular-arbuscular mycorrhizal fungi for mineral uptake. He described the magnolioid root as being generally devoid of root hairs, and many of these plant species respond to mycorrhizal infection even in relatively fertile soil. In contrast—Baylis noted—graminoid roots, which frequently have a dense cover of long root hairs, generally respond to mycorrhizal infection only in the most phosphorus-deficient soil. Baylis noted that tropical species were not used in his investigation of root hairs and mycotrophy.

In his studies in an Amazonian rain forest, St. John (1980) observed that none among the Magnoliales had root hairs, and all were either heavily or moderately infected with VA mycorrhizal fungi. Among the graminoid roots, only 15 species had root hairs, of which four were 'rare', five 'inconstant' and six 'constant'. A good many graminoid roots were completely non-mycorrhizal or only lightly infected. However, Read *et al.* (1976) had reported that members of the Gramineae (now Poaceae) were particularly heavily infected with VAM.

Khan (1974), who studied the occurrence of mycorrhiza in halophytes, hydrophytes and xerophytes, reported that mycorrhizae were found mainly in the surface and subsurface horizons of the soil, and were much less abundant in the deeper layers. Differences in root resistances between mycorrhizal and nonmycorrhizal soybeans have been reported (Safir *et al.*, 1971) and according to the authors, the external hyphae might increase the surface area of the root system much as would an increased number of root hairs and thereby enhance nutrient uptake.

b) Root hairs : VAM relationship in aquatic plants

The root system in submerged aquatic plants were earlier believed to be devoid of VAM fungi as are the root hairs (Khan, 1974; Hayman, 1978). Shuja *et al.* (1971) did not find any presence of VAM fungi in roots of *Populus euroamericana* growing towards water on the banks of a canal, while roots growing away were colonized by both ecto- and endo-mycorrhizal fungi. Maeda (1954) reported that plants growing in water or damp soil were non-mycorrhizal, but could become infected if grown under drier conditions. Mejstrik (1976) reported mycorrhiza in some common emergent plants growing in peat bogs in Czechoslovakia. Infections were dependent on water-level fluctuations and only developed when the water table dropped below ground level. Similarly, Manjunath *et al.* (1981) noted VAM infections can occur in rice under semi-aquatic conditions and that infections increased under non-flooded conditions.

However, Keeley (1980) gave the first experimental confirmation that VA mycorrhiza could tolerate submersion and showed that in Blackgum seedlings (*Nyassa sylvatica*), the highest infection occurred near main roots and decreasing outwards. This was attributed to limited oxygen transport to distal roots under flooded conditions. Later, Chaubal *et al.* (1982) recorded VA mycorrhiza in aquatic species, including *Hydrilla verticillata* (L.f.) Royle, and they related infections to temporary drying following low water levels. Most of the infected species reported were either emergent or free floating, or otherwise have been periodically exposed to semi-aquatic conditions.

High levels of mycorrhizal infection in shallow water plants, particularly those without root hairs, were reported by Clayton and Bagyaraj (1984), whereas most of the plants with abundant root hairs were not infected. This indicated that the function of root hairs is probably taken over by mycorrhizae in the absence of root hairs. However, one species (*Myriophyllum triphyllum* Orchard,) was an exception to this root hair-infection relationship, which had no infection and no apparent root hair cover. Further, Sondergaard and Laegaard (1977)—on the basis of their observations—concluded that aquatic plants without root hairs tend to be mycorrhizal, implying thereby, that the plants with profuse root hairs need not depend on the mycorrhizal association for their nutrition. But, to the present author, it appears that the functional role of VA mycorrhizal fungi and root hairs is not the same, as evident from the fact that there is no such correlation between mycorrhizal infection and the presence or absence of root hairs as root hairs were present in all but one of the plant species studied (Bagyaraj *et al.*, 1979). *Cyanotis cristata* and *Eichhornia crassipes*, which were reported to possess profuse root hairs, were found to be highly mycorrhizal, while *Cyperus eleusinoides*, which did not possess root hairs, was non-mycorrhizal. The present author reinvestigated for the presence of root hairs in the submerged roots of *Eichhornia crassipes* and it was discovered that root hairs were completely absent (**Fig. 1a**) thus confirming the observations of Charlton (1978). The roots of most of the emergent and floating-leaved aquatic plants, develop profuse root hairs, as has been pointed by Sculthorpe (1967).

Do root hairs aid in VAM infection?

There are some earlier studies to show that the VAM hyphae generally penetrated either through root epidermis or through root hairs (Mejstrik, 1972). The greater number of entry points in *Molina coerulea* rather than in *Galium verum* was quite clear cut, and probably due to several causes, including the development of root hairs which were relatively frequent in *M. coerulea* than in *G. verum*. According to Meisner and Karnok (1991), there may be a possible interaction between root hair occurrence and mycorrhizae infection.

Michelsen (1993) on studying the roots of 28 species of epiphytic vascular plants collected from tree trunks found no correlations between root hair abundance, root hair length and VAM colonization.

All these findings show that the root hair-VAM fungi relationship in aquatic and terrestrial plants is extremely intriguing and needs more research.

c) Root hairs : Ectomycorrhizal relationship

Ectomycorrhiza are also observed to infect and grow preferentially in the root hair zone and the bent tips of infected hairs are evident in *Tectona grandis* (author's own observation). However, later studies have provided explicit evidence that the initial focus of ectomycorrhizal infection is the cap region of the root, because there was specific and rapid build up of hyphae only around root caps (Horan and Chilvers, 1990). These authors feel that there is a selective chemotropic attraction of these mycorrhizal fungi to substances diffusing from compatible host root apices.

Root hairs have been observed to have involved as the entry points for the mycorrhizal fungus into the host root (*Spathoglottis plicata* Blume) and also serve as a place for the reproduction of the fungus and for their exit from the root cortex to the rhizosphere (Kumar and Krishnamurthy, 1998).

Chilvers and Pryor (1965) had reported that root hairs were completely suppressed by mycorrhizas in eucalypts. Ditengou *et al.* (2000) believe that endogenous Hypathorine counteract in controlling root hair elongation during ectomycorrhizal development and the absence of root hairs in ectomycorrhizal plants might be due in part to fungal release of molecules, such as hypaphorine that inhibit elongation of root hairs. Ericales members and orchids which lack root hairs are associated with ectomycorrhiza, which would be an advantage in the absence of root hairs (Brook, 1952; Nieuwdorp, 1972).

Peterson and Farquhar (1996), in their review on root hairs, have covered some aspects of the relationship between root hairs and mycorrhizas. According to them, although ectomycorrhizal roots characteristically lack root hairs, their absence is usually due not to an inhibition of development but due to their incorporation into the mantle. It was suggested by Hoveler as early as 1892 (cf. Baylis, 1975) that plant species without root hairs were the most consistently VA-mycorrhizal, a hypothesis that has since been corroborated by Baylis (1975), who suggested that these VA-mycorrhizal plants with few or no root hairs actually obtain greater benefits from the extra radicle hyphae than could be provided by the root hairs alone. Peterson and Farquhar (1996) believed that although this correlation between root hair absence and colonization is not understood, it is possible that there is a dependence of the plant upon the fungus for its function as root hairs. The present author attributes this association—either ecto or endo—to enhance oxygen intake into the root system, besides their incidental functions such as absorptions.

4. ROOT HAIR AND ROOT PARASITISM

Parasitism in sandal (*Santalum album*) is not well understood for want of precise knowledge in the mechanism of absorption of nutrients by the haustoria. Thus, the question arose whether the haustoria were substitute to root hairs in plants. No doubt sandal seedlings have root hairs in their roots; once the haustoria begin to form on the roots, root hairs are not to be seen. Barber (1906a) suggested that the absence of root hairs in sandal tree is indicative of the possibility of its dependence on host plants for inorganic nutrition. Baird and Riopel (1983) demonstrated that root hairs of *Agalinis purpurea* play an important role in its attachment to a host. The hairs are the first haustorial structures to contact the host surface and hosts that lacked root hairs were not found to attach to form haustoria. The Cation Exchange Capacity (CEC) is very low (7) in sandal and this perhaps explains one reason for its parasitic habit. Such hosts which exhibit rise in CEC on parasitization by sandal are considered best hosts (Kunda Deval *et al.*, 1971; Parthasarathi *et al.*, 1974).

Baird and Riopel (1985), who studied the root hairs of parasitic Scrophulariaceae members, found that normal root hairs were never observed attached to a host root, while haustorial root hairs which develop from the surface of the haustorium and have increased amount of coating material, were the first to contact the host surface. Haustoria that lacked hairs were not observed to attach to host roots (Baird and Riopel, 1984). The coating material on the haustorial hairs is believed to help in forming a continuous sheet, thereby enhancing adhesion of the haustorium to the host roots by increasing the effective area of contact. Infection of tomato roots by *Orobanche ramosa* L. was observed in the root hair zone and branching points of the roots (Hameed and Foy, 1991).

5. ROOT HAIR AND INFECTION BY NEMATODES AND PATHOGENS

Gall-forming root nematodes and root pathogens also seem to prefer the root hair region for their invasion. Orion and Lapid (1993) reported that the nematodes invaded the roots of *Vicia sativa* at the root hair region by forming a clear 'drilled' hole in the root epidermis. The surface of the invaded region appears as a lesion on which the root hairs were shed and at the lesion margin, abnormally long root hairs were observed.

Dong *et al.* (1993) have found that three strains of *Erwinia carotovora* pv. *carotovora* established a relationship between adsorption on root hair surfaces and invasion of root tissues in Chinese cabbage. Concentration of bacteria was usually highest on the tips of the root hairs. As reported by Prinsloo *et al.* (1992) in chicory roots, the germ tubes of *Thielaviopsis basicola* formed appressorium-like structures from which penetration of root hairs occurred. The infection then spreads to the tap root. Decay of the root hair zone adjacent to the uninfected stem was apparent 5 days after inoculation. *Plasmodiophora brassicae* has been observed to infect *Brassica* spp. through the root hairs (Voorrips, 1992).

6. OTHER FUNCTIONS OF ROOT HAIR

i) Role of mucigel

Several authors have tried to ascribe various other functions to the root hairs. A layer of mucilaginous material, the mucigel, present over the root hairs is presumed to improve the root's water-holding capacity by enhancing contact or adhesiveness between the root hair and soil particles (Clarke *et al.* 1979; Chaboud, 1983) and certainly providing a favourable medium for the growth of rhizosphere microorganisms (Umali-Garcia *et al.*, 1980; Berry *et al.*, 1983; Kramer, 1983). Moreover, lectins forming small dense patches on the tip of young growing hairs of *Pisum sativum* L. may allow cell-to-cell adhesion between *Rhizobium leguminosarum* and root hairs (Diaz *et al.*, 1986). Whether all plant root hairs possess this mucigel and carry out similar function is questionable. However, the primary roots of germinating pine seedlings are encased in a mucilaginous sheath that apparently functions in maintaining root hydration (Berlyn, 1967, 1972). Bhat *et al.* (1976) suggested that mucigel secreted from the root hair tips may have played a major part in the absorption of phosphorus. According to Young and Martens (1991), mucigel exuded by hypocotyl hairs and root tip cells aids in improving radicle penetration into the substrate. Greenland (1979 cf. Lamont, 1982) noted that root hairs act as links across gaps between the soil particles and root that develop as the soil dries, while Nambiar, (1976 cf. Lamont, 1982) believed that the mucigel they secrete serves to plug these gaps.

ii) Role of root hairs in drought resistance

Drought resistance in *Ferocactus acenthodes* and *Opuntia ficus-indica* was attributed primarily to the formation of soil sheaths in the root hairs zone (North and Nobel, 1992). A constant feature of the sheathed region of the grass roots, as observed by McCully (1987), is the presence of numerous root hairs, most of which have knobbly curled regions. Occasionally, some of these hairs branch at such regions. Soil particles cling tightly to the surface of the curled hairs but not to straight portions even of the same hair. The clinging nature of soil particles to distorted portions of root hairs has generally been attributed to the growing hairs being forced to bypass the particles blocking their path. At the distorted regions of the root hairs, soil particles are tightly appressed to the wall, as indicating gelatinization of the wall. According to Arbor (1934), in the sheath-forming desert grass *Aristida pungens*, the root hair-bearing epidermis secretes the mucilage, which aments the sand particles. There is no modern study of localized change in adhesiveness of the root hair surface.

In the recent review on root hairs, Hofer (1991) found it difficult to draw up an inventory of the physiological functions held by the root hairs. They remain very interesting cells to be studied and the physiological roles assumed by the very long hairs formed on some submerged hydrophytes remain to be investigated. Certainly, the importance of these tiny structures on the metabolism of the entire plant by increasing the surface of contact between the root and its surrounding is realized.

Cormack (1949, 1962), Hofer (1991) and Peterson and Farquhar (1996) have made major reviews on root hairs. Peterson and Farquhar (l.c.) dedicated the review article to the late Dr R.G.H. Cormack (deceased 1995), who had done a pioneering work on root hairs. Peterson and Farquhar (l.c.) claimed to have reviewed integrating new information on root hairs but they have omitted to include the publication of Bhaskar *et al.* (1993) who for the first time propounded the theory of the 'respiratory role' being the chief function of root hairs.

Root Hairs as 'Respiratory Gills': Evidences

1. ROOT RESPIRATION

a) How do roots get oxygen?

The possibility that roots may obtain oxygen from the atmosphere not only via the soil but also internally via the aerial parts has been considered for some time (Greenwood, 1967). Contradicting this theory, Healy and Armstrong (1972) proposed that for root growth of an initial length (8-9 cm), internally supplied oxygen will meet all respiratory requirements. By means of ^{15}O tracing, the rapid internal diffusion of oxygen from the foliage to the roots has been directly demonstrated in broad bean (*Vicia faba*) seedlings (Evans and Ebert, 1960) and in barley and rice seedlings (Barber et al., 1962). After growth to a certain length-including the formation of laterals, a more static aerated medium will be required in order to prolong the growth. The accumulation of a larger respiratory sink in the more extensive root system may mean that internal oxygen supply becomes inadequate to the remote parts of the root system. Therefore, it becomes necessary for the extensive root system to get more quicker oxygen supply from soil or air for continuous growth. Plants might be able to survive soil anaerobiosis or root anaerobiosis by increasing the number of adventitious roots, which would have higher root porosities. Internal oxygen supply in mesophytes might be adequate for minimum plant responses and metabolic activities, but still inadequate for maximum growth requirements. [Reduction in growth rates and dry weights for all plants has been observed in full flooded as compared to non-flooded treatments (Yu et al., 1969).] Therefore, where from do the grown up roots get oxygen?

b) Where does oxygen enter roots?

It is well known that roots exhibit salt respiration in relation to ion uptake, both anions and cations (Lundegardh, 1955; Jennings, 1986), and respiration of roots is the single largest contributor to soil CO_2 evolution (Ewell et al., 1987). But whether all the different tissues in root system take equal part in active respiration or growth respiration is not clearly understood.

Hardly any reports exist on the respiratory rates of different segments along the root. Machilis (1944) is perhaps the earliest study which indicated that the rate of respiration (O_2 uptake) of the apical 10 mm root segment of *Hordeum vulgare* was several times higher than that of older segments along the root. But the respiratory quotients of the different regions of the root were equal. According to Yemm (1965) the root meristem has relatively low respiration rate on a cell basis. Upon vacuolation there is a marked rise in the rate per cell. Beyond this zone, respiration is maintained at a high level. In maturing cells, respiration per cell is somewhat lower again, but still about 5 times higher than in the meristematic cells. Despite the high energy required in the root tip, the alternative path still contributed significantly in the 5 mm tip of *Zea mays* roots. Subterranean roots have a thin cuticle and no stomata, but may develop thick cuticle and stomata when exposed to sunlight (Dawes and Bowler, 1959). Root hairs are the only cells in the roots which lack cuticle. The presence of cuticle on the other parts of roots would prevent diffusion of any gases, but the root hairs may permit since they lack cuticle. But this was interpreted by the plant physiologists as a structural advantage for water and mineral absorption.

Mathematical models

While it has not proved possible to determine the respiratory rates for individual root tissues in vivo, differences between the various tissues are certain to exist and several mathematical models have been designed to accommodate such differences (van Noordwijk and de Willigen, 1984; de Willigen and van Noordwijk, 1984; Armstrong and Beckett, 1985). Treating the root as a homogeneous cylinder with respect to oxygen diffusibility is a mathematically-convenient assumption that has long been an inherent weakness and potential source of error in modelling the oxygen relation of soil-aerated roots (Armstrong and Beckett, 1985). van Noordwijk and de Willigen (1984) hypothesized that only a part of the root circumference is in contact with the water continuum of the soil and hence this probably restricts the possibilities for oxygen uptake to the remaining part of the circumference because the water film has often been considered as being the major obstacle for oxygen diffusion into the roots (Lemon, 1962). Armstrong and Beckett (1985) predicted that the root wall layers, including hypodermis, may impede diffusion of oxygen into the cortex. Thus, thin root walls will be particularly advantageous.

Vartapetyan and Jackson (1997), after intensive research in the field of plant adaptations to anaerobic stress (oxygen shortage affecting roots has been well emphasized), have reported on the morphological and physiological escape mechanisms including formation of replacement roots through adventitious rooting at the shoot base, aerenchyma development and internal aeration pathways. However, the role of root hairs in enhacing oxygen diffusion into the roots and prevention of anaerobiosis has not been reported till date.

The present author has attempted here to show through various experimental and other evidences that root hairs are specialized to serve as the main avenues for oxygen diffusion in the roots, and possibly complementary with various absorptive functions. The present theory presents an inventory of evidences that root hairs are specialized chiefly for intake of O_2 and respiration, which is intimately involved in the cooperative uptake of water and minerals.

2. ROOT HAIR AS AN ENTRY POINT FOR OXYGEN DIFFUSION

It can be amply supported with evidences that the important function of root hairs is to aid in oxygen diffusion into the roots serving the same purpose as stomata and lenticels in aerial shoot. The root hairs act as short circuits to the soil air, permitting greater respiration rate in root hairs and relatively speedier and greater transport of ATP or O_2 or both to the underground organs than would be possible if supplies depended entirely on diffusion from the aerial foliage through a long route. Occurrence of unthickened or thinned areas of cell walls, presence of a thin or no cuticle and dense cytoplasm studded with mitochondria, are some of the special ultra-structural adaptations of root hairs which offer an evidence that at least one of the main functions of root hairs is to act as the entry point for O_2 diffusion and respiration.

The present theory that root hairs serve as 'gills' in roots is also based on six additional pieces of circumstantial evidences: (1) development of denser root hairs in soil voids; (2) development of dense root hairs on the juvenile primary root of viviparously germinating seeds inside pods; (3) production of new adventitious roots densely clothed with root hairs outside the soil exposed to the air when the soil is saturated with irrigation; (4) occurrence of permanently open stomata in the root hair zone of some species which may be a double advantage as it may enable additional diffusion of oxygen to meet the greater energy requirement of growing seedlings; (5) occurrence of well developed root hairs in most of the terrestrial plants and their absence or poorly developed nature in hydrophytes; and (6) development of maximum number and length of root hairs in relatively dry and porous soils rather than in the wet soils.

It is postulated that root hairs are produced chiefly in soil voids which contain air to increase the surface area enabling maximum diffusion of oxygen, thereby protecting the below-ground roots which usually suffer from anaerobic stress.

3. EVIDENCES DISFAVOURING WATER AND MINERAL UPTAKE AS THE MAIN FUNCTIONS OF ROOT HAIR

The subterranean parts of higher plants have been studied to a lesser extent as compared to above-ground parts and, consequently, their physiological functions are less well understood. Root hairs are a part of the underground system whose functions remain interesting as they constitute a very small portion of the root system.

Various functions have been ascribed to the root hairs as per the accumulated literature. Among them, absorption of water and ion uptake are the important ones. They are generally considered to increase the surface area of roots and thereby increase absorption of water and mineral nutrients from the soil (Dittmer 1937; Rosene 1943; Cailloux 1943; Mackay and Barber 1985, 1987). However, the structural and functional mechanisms by which this is accomplished remains in doubt.

That root hairs absorb water was first negated by Coupin (1919), a theory that shook all the previous theories. Coupin later modified his previous statement that root hairs are not the main organs of absorption. According to Russell (1977) and Drew (1987), the above types of observations do not by themselves establish that root hairs play a major role in absorption. These authors concluded that there are no grounds for ascribing a specialized role in absorption to root hairs. The current consensus is that water uptake by roots occurs just by passive means and thus root hairs have no specialized function in water uptake. Dittmer (1937) and Mackay and Barber (1985, 1987) argued that root hairs increase the surface area of the roots several fold and thus, just by surface dimension alone, constitute an important component of the water uptake system of roots (Dittmer 1937; Mackay and Barber, 1985, 1987). The assumption that greater the surface area, greater the absorption of water and nutrients is not quite correct. Even though the total surface area of root is greater (as great as 22 times) than the aboveground transpirational area, the actual water absorption area is confined to the root hair tips. The old root hairs may not take part in water absorption as they will have lost living protoplasm. Therefore, presence of persistent root hairs as in *Gleditsia* (Whitaker, 1923), *Ailanthus malabarica* (Author's observation) may not have any absorptive significance.

Other work on the role of root hairs in the absorption of water has been done but the adaptive significance of root hairs in water absorption remains unclear (Hayward *et al.*, 1942; Sierp and Brewig, 1935; Clarkson, 1981; Clarkson and Hanson, 1980; Kramer, 1983; Davies, 1986; Clarkson *et al.*, 1987; Jones *et al.*, 1983; Kurkova, 1981).

There are some studies indicating that most water absorption takes place preferentially in the younger roots and specifically in the root hair zone (Devlin and Witham, 1983). Thus, root hairs assume importance in water uptake beyond merely increasing their absorptive surface. The reason for this selectivity is assumed to be the lack of suberized tissues in the root hair zone, implying that much of the plant's root system does not absorb much water. However, Addoms (1946) observed that in yellow poplar (*Lyriodendron tulipifera*), Sweet gum (*Liquidambar styraciflua*) and short leaf pine (*Pinus echinata*), the suberized roots absorbed a dye solution. He further pointed out that there are three portals of entry for water through such suberized roots: lenticels, breaks around branch roots and wounds (cracks). Hayward *et al.* (1942) showed that maximum absorption of water takes place 10 cm behind the root tips in roots over 7 cm long in onion and decreases toward both the tip and the base. On the contrary, Sierp and Brewig (1935) had found that although in *Vicia faba*, the maximum intake of water occurred 1.5-8.0 cm behind the tip in a root over 10 cm when the rate of transpiration was increased, the absorbing zone was extended and the region where more rapid absorption occurred shifted toward the base of the root.

Studies with potometers have revealed that a little water enters through the meristematic region in the root tip because of the high resistance offered by the dense protoplasm and lack of xylem elements to carry it away. Further back xylem is functional, but suberization and lignification of the hypodermis and especially of the endodermis seriously reduced the entrance of water and minerals (Kramer, 1983). According to Davies (1986), most of the water will flow along the path of least resistance. As roots get older, their permeability declines, but in woody perennials possessing only a comparatively small percentage of unsuberized roots, considerable absorption of water must occur through older suberized roots where the epidermis and even the entire cortex may be replaced by periderm and secondary phloem (rhytidome). Studies by Clarkson et al., (1987) in *Zea mays* have shown that water uptake by root axes, as measured by micropotometry, was greater at distances more than 100 mm from the root tip than in the apical zone where the hypodermis was unsuberized. In the more mature zones of roots, the rate of water uptake was found to be greater even though hypodermal suberization was more marked. Recently, Kramer and Boyer (1995) have provided a complete description of the evidence for water absorption through suberized roots. The evidence presented includes nuclear magnetic resonance images of water depletion around woody and suberized roots. With this it becomes clear that water absorption is not the chief function of root hairs and they are not specially organized for the purpose of water obsorption and also that the plant roots can continue to absorb water even in the absence of root hairs.

Pressure probe tests have shown that the hydraulic resistance of root hairs is similar to that of hairless epidermal cells and cortical cells (Jones *et al.*, 1983), which does not suggest either any specialization of epidermal plasma membranes or greater plasmodesmatal frequency between root hair cells and the cortex (Drew, 1987). Studies on the function of root hairs and adjoining hairless epidermal cells in the higher aquatic plant (*Trianea bogotensis*) have shown that net K^+ influx was virtually restricted to the root hairs. Root hairs represent a major uptake site for macronutrients from the soil, but the molecular mechanism of the transmembrane uptake remains largely unknown (Gassmann and Schroeder, 1933). Clarkson (1981) and Clarkson and Hanson (1980) also reported that older regions of roots can take up mineral nutrients. Drew (1987) observed that in radish (*Raphanus sativus*), there was no difference in the number of plasmodesmata between root hair cells and hairless cells. So, according to him, there is no ground for ascribing a specialized absorptive function to root hairs in general. Dr K.G. Raghothama, Professor and University Faculty Scholar, Centre for Plant Stress Environmental Physiology, Departmental of Horticulture, Purdue University, USA, in his personal communication to the author (26 July 2002), mentioned that phosphate tranporters are enriched in the whole root epidermis of Pi deficient roots, including root hairs. He also confirmed that this is also indicative of the involvement of whole root epidermis in phosphate uptake, in addition to root hairs. Thus, root hairs are not essentially required for mineral uptake but are necessary for some other purpose, which will be discussed later in this chapter.

Several authors have tried to ascribe various other functions to the root hairs. A layer of mucilaginous material—the mucigel—present over the root hairs is presumed to improve water holding capacity by enhancing the contact or adhesiveness between the root hair and soil particles (Clarke *et al.*, 1979; Chaboud, 1983) and certainly provide a favourable medium for the growth of rhizosphere microorganism (Umali-Garcia *et al.*, 1980; Kramer, 1983). Whether all plant root hairs possess this mucigel and carry out similar function is questionable. However, the primary roots of germinating pine seedlings are encased in a mucilaginous sheath that apparently functions in maintaining root hydration (Berlyn, 1967, 1972).

The role of root hairs in the entry of nitrogen-fixing microorganisms into the root systems, as in the case of N_2-fixing leguminous plants, is very well established. Although exceptions exist, the importance of root hairs for nodulation in most leguminous species is evident from the complete absence of root nodules in groundnut mutant lines, which also lack root hairs (Nambiar *et al.*, 1983). The situation in actinorrhizal plants is more complex because some species are obligatorily infected by *Frankia* through the root hairs, whereas others may be infected through the epidermal cell walls and still others by a combination of root hairs and epidermal cell walls (Baker, personal communication). In case of mycorrhizal plants, in the absence of root hairs, the function of root hairs is said to be taken over by mycorrhizae (Baylis, 1975; St. John, 1980; Clayton and Bagyaraj, 1984;

Hofer, 1991), but in few exceptions, plants with dense root hairs also exhibited VAM infection (Clayton and Bagyaraj, 1984).

It is evident from the above information that the physiological role of root hairs in plants is not fully understood. No doubt root hairs have different specialized functions in different plant groups, like root nodulation in legumes and actinorrhizal species, and mycorrhizal associations of roots, but this is not the primary function of root hairs in non-nodulating and non-mycorrhizal plants.

4. EVIDENCES IN SUPPORT OF THE RESPIRATORY ROLE AS THE CHIEF FUNCTION OF ROOT HAIR

a) Concentration of hairy roots in the soil surface as evidence of their role as 'respiratory gills'

The fine bunches of hairy roots of plants are usually greatly concentrated in the upper soil horizons having closer proximity to an oxygen source. In *Kingia australis*, small laterals bearing root hairs proliferate among the persistent leaf bases (Lamont, 1982). Proteoid roots, which are characteristic of Proteaceae (Lamont, 1982) and which occur as root clusters or bunches of hairy roots, are concentrated in the uppermost 10 cm of the soil. They often form a dense mat at the soil surface (Lamont, 1982). This mat is usually regarded as a device for trapping nutrients. Contrary to the belief that their main function is water and nutrient uptake, it can now be stated that these metabolic processes take place mainly in the availability of O_2 and the energy produced by respiration.

b) Absence of root hairs in roots grown in tissue culture medium

Normally, root hairs do not grow on roots that develop in a tissue culture medium, most likely because of poor aeration or non-availability of free O_2 in the liquid medium. In tissue culture propagated plants (Cashewnut, banana and *Vanda*), as observed by the present author, only those roots that grew outside the medium or away from the culture medium possessed dense root hairs while no root hairs were observed on roots submerged in the medium (**Fig. 11**) with the exception of *Solanum tuberosum*, which showed root hair development even inside the medium. Nambiar *et al.* (1983) observed that the root hairs were produced only above the nutrient solution level in *Arachis hypogea*. Studies of Yie and Liah (1977) have also shown that root hairs do not grow on roots that develop in agar. Roots bearing root hairs are produced, when shoots are cultured in liquid media and are formed in still greater numbers when the shoots are supported above the medium on filter paper bridges. Abo El-Nil and Hildebrandt (1971), Mascarenhas *et al.* (1978), Harris and Stevenson (1979), and Adejare and Coutts (1981) reported that roots bearing root hairs were produced *in vitro* when cassava shoots were dipped into a commercial rooting powder before being placed on a solidified culture medium. According to Debergh and Maene (1981), *in vitro* roots frequently lack root hairs. The oxygen concentration in the culture vials covered with cotton plugs will be closest to that in the ambient atmosphere and the *in vitro* root, which stands free of the medium and is surrounded by a minimum film of moisture or the medium gets good aeration. Submerged tissues or organs in a static medium are very poorly aerated. With unventilated cultures growing in sealed vessels, oxygen concentration at the level of the medium or the roots can be considerably less than the same found externally. This is due to the fact that use of oxygen by the culture creates a local deficit, which may not be immediately compensated because of the impedance to diffusion created by closures, especially if they are tightly fitting and impermeable. Kozai *et al.* (1988) found that there were 0.1 air changes per hour inside 47 ml tubes covered with aluminium caps, 1.0 change when they were plugged with plastic caps, but 6.2 changes per hour if the tubes were covered with micro-porous polypropylene.

Tomato roots have been found to grow more rapidly if a liquid medium is subjected to a continuous gentle agitation (Said and Murashige, 1979). Roots initiated on the shoots of some plants can be seen to grow on the surface of agar with a more dense cover of root hairs, rather than down into it, suggesting that oxygenation of the medium is limiting. Oxygen is only sparingly soluble in water and more so in incubation temperatures from 21°C to 25°C and in a plant culture medium. Dissolved salts and non-electrolytes such as sucrose further diminish the solubility of oxygen in the medium.

Bordonaro and Curtis (2000) developed an experimental system that produces hairy root cultures of *Hyoscyamus muticus* with and without profuse root hairs. To assess the impact on bioreactor performance, hairy and hairless root cultures were grown in a highly characterized 15-L bubble column bioreactor. In the absence of root hairs, the mixing was greatly enhanced. The results showed that the root hairs, which are said to facilitate nutrient uptake in a static environment, are detrimental to growth in a liquid environment as an effect of their stagnating fluid flow and limiting oxygen availability.

Hence, adequate gaseous exchange is essential to ensure root hair development, adequate oxygen diffusion for ATP production which are prerequisites for a normal growth and development or fast rate of multiplication in most tissue cultures.

c) Development of root hairs in soil voids and aerobic environment

Root hairs are often most conspicuously developed on roots growing in soil voids of appreciable size (Russell, 1977). Further, more dense root hairs have been found to develop in relatively dry and porous soils than in too wet soil (Pfeffer, 1897 cf. Whitaker, 1923; Bhat and Nye, 1973; Mackay and Barber, 1985, 1987). Root hairs of orange tree seedlings grown in vermiculite seemed to be longer than in seedlings grown in paddy soil and granite soil (Poerwanto *et al.*, 1987). The role of oxygen in root hair formation is also evident from the fact that most roots generally produce root hairs much more abundantly in moist air than when submerged in water. An optimal level of humidity seems to be one of the essential factors for the development of the root hair (Uphof, 1962). Seed germination in the humid air increased the percentage of radicles with root hairs and increased the length of both the root hair zone and the longest root hair (Pill *et al.*, 1987).

The present author conducted a simple experiment by placing finger millet and rice seeds for germination in petridishes (**Fig. 7**). In one treatment (aerobic), the blotter paper was made to become just moist and seeds were placed over the moist paper. In the other treatment (anaerobic) excess water was added so as to submerge the seeds in water. In the aerobic condition, the radicles developed very dense and long root hairs and were directly exposed to air rather than to water, while in the anaerobic condition, the seedling roots were completely bare of root hairs. However, hairs had developed only in those portions of roots exposed to air. In the aerobic condition, seedlings grew normal and faster than under anaerobic condition (**Fig. 7**). In onion also, it was noticed that roots submerged in water did not develop root hairs while parts of roots which ought to have developed root hairs and which are exposed to air do develop root hairs. Further, when the water level was brought down, the root parts which had become exposed to air developed root hairs subsequently.

The present author has observed that in an excessively irrigated tomato plot, roots with whitish dense root hairs had grown out of the soil while the root hairs inside the soil had obliterated or possessed no root hairs. Moisture was not a limiting factor for these cases, but still root hairs developed on exposed parts, indicating that root hairs are not meant to increase water absorbing area but to increase oxygen supply to the root tissues for root respiration. The soil was saturated with irrigation and in order to survive from O_2 stress, fresh roots were produced towards the gaseous oxygen source, i.e. air.

The development of root hairs is influenced by such soil physical characters which allow proper soil aeration. More dense root hairs have been reported to develop in porous soils (Bhat and Nye, 1973; Mackay and Barber, 1985, 1987). Teak seedlings cultured in clayey or heavy soil did not develop root hairs. On the contrary, when they were cultured in coarse sand, all the newly-induced primary roots produced dense root hairs (Bhaskar, unpublished). All these evidences prove that root hairs are promoted to develop only in a gaseous environment and not in heavy and saturated soils.

Further, the present author has also observed in fresh pea (*Pisum sativum*) seeds germinating inside the pod having dense root hairs on the radicle growing into the air spaces inside the pod (**Figs. 6a** and **b**). *Phaseolus vulgaris* seedlings grown only on a damp paper towel also showed an enhanced presence of root hairs on the radicle above the paper towel. In both of these experimental systems, the mineral nutrition necessary for growth came from the cotyledons. If the special respiratory role for root hairs was incorrect, one would not expect these root hairs to be formed in seedlings developed inside the pod where there is no liquid medium. These fast-developing seedlings require sufficient quantity of energy in the form of ATP. The presence of dense root hairs with consequent increased surface area would enable a sufficient amount of oxygen diffusion through root hairs to break down stored food material and supply energy to the growing seedling. In a sense, this profusion of root hairs evidently serves as a 'root gill'.

Some results with experiments on rice are interesting. Under normal conditions, radicle protrusion is the common mode of germination in rice and nearly all other species exposed to air. Under hypoxia or anoxia, rice seeds germinate by upward protrusion of the coleoptile rather than protrusion of the radicle. Root hardly develops, but the coleoptile continues to grow under hypoxia (Atwell and Greenway, 1987; Atwell *et al.*, 1985; Alpi and Beevers, 1983). In spite of detailed biochemical work, it was considered an enigma as to explain how rice plants can tolerate hypoxic conditions unusually well. But this can be explained. Oxygen or air seems to be crucial for the stimulation of protrusion of radicle. First, the coleoptile develops into shoot and as the leaves with stomates are developed, it allows oxygen to enter into the lacunar system in the stem. Once the oxygen supply is ensured, the radicle starts its further development. If the radicle was to develop first under a submerged condition, the seedling would soon die for want of oxygen. This is how the rice plants or any similar plants seem to have adapted to survive under a submerged or flooded condition. The relationship between reduced O_2 supply in roots and the stimulation of aerenchyma has been explained in terms of increased ethelene synthesis (Justin and Armstrong, 1991).

Root hairs are highly sensitive to direct exposure to air when they lose their turgidity, twist and collapse or sometimes excrete. In the absence of oxygen or by the addition of malonate at concentrations that inhibit respiration, the root hairs tend to excrete. It is perhaps possible that maximum diffusion of gaseous oxygen or oxygen dissolved in moist air present in soil voids, takes place into the root hair cytoplasm and this plays a predominant role in increasing its degree of oxygen saturation in root. The exchange of atmospheric oxygen across the air-water interface has been found 2.5 to 3.5 times the amount liberated during primary production (Manna et al., 1992).

Adventitious roots of *Kalanchoe fedtschenkoi* produce multicellular hairs (doubtfully root hairs!) in air but stop producing them when the roots penetrate the soil (Popham and Henry, 1955). In certain hydrophytes and marsh plants such as *Ludwigia repens* L., white cellular floats are present at the nodes while in *Ludwigia octovalis* Sw., the submerged roots produce some roots which grow above the surface of water and this aerial part of root is densely covered with whitish root hairs while the root just below it inside the water is bare of hairs. Therefore, the proliferation of root hairs in air would not enhance either water or mineral nutrient absorption.

Pill and Beers (1987) studied the effect of germination in humid air and aerated water on root hair frequency in tomato seeds. They observed that seed germination in humid air increased the percentage of radicles with root hairs and increased the length of both the root hair zone and the root hair. This shows that in humid air, more root hairs are produced in order to increase the surface area to get the required amount of oxygen.

Of late, aeroponics is gaining more importance in contrast to hydroponics in the commercial cultivation of vegetable crops. Aeroponics is a method of growing plants by delivering a nutrient mist to the roots using hydro-atomizing spray jet (misting nozzle). The nutrient solution is released at high pressure, producing a nutrient mist. The main advantage of aeroponics has been the induction of large amounts of well developed root hairs (in contrast to hydroponics), thereby well oxygenating the roots.

d) Cytochemical assay prove elevated respiratory activity in root hairs

Cytochemical assay to localize respiratory activity in hairy root submerged cultures has shown that when treated with nitroblue tetrazolium, the hairy root (Agrobacterium rhizogenes infected) cultures of *Beta vulgaris*, *Nicotiana tabacum* and others showed rapid and intense staining at the apical meristem and lateral root initials; which is indicative of elevated respiratory activity in these regions (Usha *et al.*, 2000). The root hairs did not show any staining, indicating their limited role in the overall respiratory budget of hairy roots grown as submerged cultures. This is a clear indication that under submerged conditions, root hairs are choked from the entry of oxygen and respiratory symptoms become feeble. But a separate study in the same line without a condition of submergence would have indicated the relative respiratory roles of different parts of root, including root hairs.

After the new hypothesis was put forward by Bhaskar *et al.* (1993) that root hairs chiefly function as 'respiratory gills' in the roots, the first experimental supporting evidence came from the cytochemical studies by Connolly and Berlyn (1996). They demonstrated a partitioning of respiration within root tissues using nitro blue tetrazolium staining and an incident light optical system that permits detailed observations of intact roots. In 4-week-old bean plants (*Phaseolus vulgaris*), deep blue formazan stain could be observed in root hairs within 2 min of exposure to the NBT mixture. Root nodules in *P. vulgaris* began to stain pink after 10 min and root apical meristems began to stain after a 20-min incubation in the reaction mixture. The remainder of the 4-week-old roots did not stain, regardless of the incubation time. In case of nodulating genera (*Phaseolus*, *Trifolium*), the root hairs were stained first, followed by root apical meristem and small root nodules. Large root nodules stained first in the portion of the nodules distal to the root proper. Thus, the root hairs in nodulating genera have been shown to possess enhanced respiratory activity. These results are again in conformity with further prediction of the author that in nodules, O_2 is required in large amounts for respiration (as the nodules respire at about 4 times the rate of rest of roots). Root hairs being the respiratory centres, rhizobia conveniently invade through root hairs. Connolly and Berlyn (1996) related the enhanced respiration activity in the root hairs of leguminous plants in the vegetative phase to Rhizobium/plant interaction involved in nodule establishment. Thus, their results showed that root hairs are either sites of enhanced respiration in some species or in developmental or physiological states of plants.

The present author conducted a chemical assay in 3-day-old seedlings of *Pisum sativum* grown inside petridishes, in order to fix the differentially active respiratory parts in a root. Germinated seedlings were placed on moist filter paper but the root hairs were exposed to ambient air inside the petridish. The seedlings were grown at ambient light-dark cycles at 29°C to 20°C day and night temperatures, respectively during the month of July 2002. The seedlings were subjected to triphenyl tetrazolium chloride (TTZC) test during forenoon. The seedlings were immersed in 1 per cent TTZC and kept for observation for differential staining of various parts of root including root hairs. In about 15 minutes, it was observed under a microscope (Olympus binocular dissection microscope) that the root tip was the first part in the root to turn red, soon followed by individual root hairs in the entire root hair zone (but not the axial part of the root) within a span of 20 minutes. This differential staining with TTZC shows that root tip being an actively growing meristematic region, exhibits greater respiration followed by individual root hairs in the root hair zone.

A similar treatment with TTZC was undertaken in the roots of freshly-germinated seeds of *Oryza sativa* (paddy), *Eleusine corocana* (finger millet) and *Zea mays* (maize). An interesting phenomenon was observed by the present author in the case of paddy seedlings treated with TTZC. Root tips and the root hairs on the aerobically-grown roots of paddy did not stain for TTZC treatment indicating as if Kreb's cycle is not taking place, i.e. H^+ ions are not realeased and seemingly respiration is restricted to only glycolysis wherein gaseous oxygen is not involved. Just to crosscheck whether the same phenomenon was occurring in other cereals, seedling roots of finger millet and maize were also treated with TTZC solution, but in these cases, the root tips and root hairs were stained as in pea, a dicot plant. Paddy being a semi-aquatic plant seems to be physiologically adopted for non-aerobic respiration in roots despite growing under an aerobic environment. This interesting phenomenon, as observed in paddy, should be further probed which might reveal interesting results.

An improved method is devised by Bidel *et al.* (2001) to measure spatial variation in root respiration using an oxygen microelectrode. The local root respiration rates were estimated by fitting models for radial oxygen diffusion and consumption to the measured oxygen profile data obtained with the microelectrodes. This technique was used in the taproot of *Prunus persica,* which showed that the respiration rate was at a maximum in the elongating apical zone. It then decreased with distance from the root tip up to the root hair zone, where it increased.

The above studies clearly indicate that with regard to respiratory activity, the root tip is the most active part, followed by root hair zone or the root hairs themselves. The root tip usually accumulates far more starch just before onset of growth and is rapidly degraded as the root tip begins to grow. Respiration is necessary to break down the stored starch in the root tip in order to produce useable form of energy. To do so, the root tip needs oxygen. It is possible that either the oxygen diffused through root hairs is translocated to the growing tip or the ATPs formed inside the respiring root hairs are translocated to the growing tip. The possibility of such a symplastic movement of oxygen and ATPs between adjoining tissues is explained elsewhere in the book.

e) Ultrastructural and histochemical evidences

The oxygen diffusion coefficient for cytoplasm is considered to be close to that of water (Tyree, 1970), suggesting that the diffusion of oxygen should be greater towards the base of the root hair which is close to the vacuole and has very little cytoplasm. Dawes and Bowler (1959) reported the presence of 'pores' (unthickened areas in the cellulose wall with thin microfibrillar network) throughout the entire wall of both young and old root hairs. These were found to increase in number and size near the base of the hair.

The overall root respiration rates are found to correlate with the number of mitochondria present in the cells (Johnson-Fanagan and Owens, 1986). The protoplasm in the root hair tip is shown to be densely packed with cytoplasmic organelles, especially mitochondria with maximum cristae (Sievers, 1963a, b). Cytochrome oxidase present in mitochondria has a high affinity for oxygen. It can function even when the O_2 concentration around it is only 0.05% of that in air (Drew, 1979). Mitochondria being represented in less numbers in non-root hair zones, diffusion of O_2 from air to cytochrome oxidase is probably retarded enough to slow the respiration rates so as to cause primarily anaerobic respiration. Since the deeper tissues in the rest of the living root cannot receive oxygen directly, they receive only dissolved O_2 or ATPs transported from root hairs and this hypothesis needs further confirmation. Tightly-packed tissues such as xylem parenchyma, and meristematic regions, which appear to have no access to air spaces, will probably be supplied with dissolved O_2 molecules or ATPs from root hairs, otherwise the diffusion of O_2 to be carried out at these parts cannot be explained. The more rapid ATP or dissolved O_2 transport system can be expected for plant roots with no lungs or haemoglobin to help transport the oxygen.

The greatest density of plasmodesmata was recorded in the wall between the root hair and the first cortical cell, and in the radial walls between the root hair and the other hairless cells (Vakhmistrov and Kurkova, 1979). Such a concentration of plasmodesmata enhances the exchange possibilities of ATPs or dissolved oxygen from the hairs to the inner tissues. Further, in simple diffusion, a type of passive transport, a molecule crosses a membrane unaided by a transport protein; gases such as O_2 and CO_2 and small relatively hydrophobic molecules such as ethanol, can cross phospholipid bilayers in this manner (Darnell et al., 1990). All these findings provide additional supports for the active respiratory role of root hairs.

Potentially, root hairs could serve the same function that the aerenchyma systems of hydrophytes serves by increasing the surface area exposed to oxygen through their penetration of soil voids where oxygen is most available. They would then serve as oxygen conduits to the rest of the root, supplying the oxygen necessary to generate ATP. Alternatively, root hairs could increase the surface area exposure to oxygen but serve as specialized respiratory cells which use all incoming O_2 as the terminal electron acceptor, reducing it to water and generating ATP which can be transported symplastically to portions of the root deficient in ATP. Such transport is not unreasonable given the amount of cytoplasmic streaming that can be observed in root hairs (Lew, 1991). Preliminary experiments conducted in Greeley laboratory, Yale University, USA suggest that such ATP transport could be operative. Over short distances, the symplastic transport of ATP seems reasonable for three reasons. Firstly, cytoplasmic streaming is observable in root hairs, and our knowledge about the trafficking of vesicles and material (still in its infancy) shows rapid and complex movement. Secondly, it would probably be better to think of the ATP not as individual ATP molecules trafficking symplastically, but rather as waves of ATP and ADP waxing and waning. Just as a candle flame is a reaction front between the paraffin wax vapour and its combination with oxygen, so can the ATP wave be thought of as a dynamic reaction front that flows and ebbs. It is also analogous to the dynamic instability of microtubules within cells. Thirdly, assuming that the function of the 'root gill' observed in *Pisum sativum* or *Jussiaea* (*Oenothera*) is to obtain a terminal electron acceptor in oxygen, the ATP must move into the developing seedling. ATP transport to cells deeper within the root can also be facilitated by root hair origins (Bogar and Smith, 1965).

Physiological and ultra-structural studies of cellular changes in plant roots under flooding (Anaerobiosis) have primarily concerned mitochondria. While attempting to correlate molecular responses with ultrastructural changes in corn roots anoxic for 8-26 hours, Aldrich *et al.* (1985) observed profound cellular effects in the first 2 mm of tissues which included proliferation of ER, production of numerous vesicles by the Golgi apparatus and long branched mitochondria. Oliveira (1977) observed mitochondrial swelling. Opik

(1973) found that rice coleoptiles possessed decreased cytochrome oxidase activity and fewer mitochondrial cristae under anaerobic conditions. Vartapetyan *et al.* (1977, 1978) and Vartapetyan (1983) also observed similar mitochondrial changes under anoxia, suggesting them to be a result of carbon starvation rather than anoxia. Aldrich *et al.* (1985) and Oliveira (1977) have noted that the portion of the root just above the meristem gives the most consistent and reproducible results. This indirectly implicates that the root hair zone as an important respiratory region in the entire root system.

Kennedy *et al.* (1990) find that the mitochondria of barnyard grass, an important weed on rice fields, even continue to function under anaerobic conditions and Vartapetian (1990) describes ultrastructural changes in mitochondria under anoxia. In an account of the 'molecular response' to anoxic stress in maize, Sachs (1990) describes the synthesis of 'anaerobic polypeptides', involving the regulation of both transcription and translocation when the cell is deprived of oxygen.

f) Soil moisture and root hairs (anaerobic environment and root hairs)

Cormack (1949) and Filippenko (1981) found that when *Zea mays* roots were grown in water, no root hairs were observed. Bhat and Nye (1973) and Mackay and Barber (1985, 1987) observed soil moisture to be the most dominant factor affecting the length and density of root hairs in maize. Under dry soil conditions, the percentage of total root length with root hairs and the density of root hairs were 1.4 to 1.8 and 2.0 to 4.1 times greater, respectively, than under wetland conditions. They averred that this may be a response to stress by the plant and helps to maintain liquid continuity around the growing root by providing greater surface area. But the present author believes that very wet soils may prevent the development of root hairs because these conditions are similar to anaerobic condition which are unfavourable for root hair development and consequently respiration.

Poerwanto *et al.* (1987) who grew the 1-year-old seedlings of *Citrus* in paddy soil, granite soil and vermiculite, found that root hairs of the trees grown in vermiculite (which obviously had more air space) were longer. In perennial ryegrass (*Lolium perenne*), proliferation of coleorhiza hairs (analogous to root hairs) was enhanced by the increased distance between the coleorhiza and water substrate (Debaene-Gill *et al.*, 1944). Under increasing dry soil conditions, the roots will tend to favour their respiratory activity by making use of greater amounts of oxygen available and hence develop more dense and longer root hairs. In this case, energy released from substrate by root hair respiration can be used to form more roots and this, in turn, increases the water harvesting capacity of the plant. It is commonly observed that there is increased allocation of carbon to roots under dry or low nutrient conditions. Old root hairs tend to excrete water, the excretion rate being faster when the root hair metabolism is lowered by the addition of malonate at concentrations that inhibit respiration (Cailloux, 1974).

Formation of special adventitious roots (stress roots) as a response to flooding may be considered to be one of the important adaptations of certain flood-tolerant species to maintain respiratory function. Many of the typical land plants may succumb to flooding because they do not produce such special roots or increase the respiratory area and obtain sufficient oxygen (Kramer and Jackson, 1954). Kordan (1976) observed in rice seedlings that an oxygen-deficient medium initiated adventitious root formation, while the aerated medium triggered the growth of roots. Topa and McLeod (1986) found adventitious roots and stem lenticels in pond pine (*Pinus serotina*) seedlings grown in anaerobic solution culture. *Eucalyptus grandis* and *Melaleuca quinquenervia* are flood-tolerant tree species and are reported to develop adventitious roots from the upper portion of tap root, lateral roots and submerged portions of stem. These special roots are fine and fibrous in nature and grow upwardly, becoming largely confined to the surface soil layer or water above the soil layer (Clemens *et al.*, 1978; Sena Gomes and Kozlowski, 1980). In contrast, certain flood tolerant species have the capacity to control anaerobic respiration in roots (Crawford, 1971, 1975, 1976; Erdmann *et al.*, 1986a, b; Jackson *et al.*, 1985; Vartapetyan, 1973; Pezeshki, 1991).

On the aspect of root-shoot relationships of flooded plants, there are two models explained by Armstrong *et al.* (1990). One of these addresses the old problem of modeling root aeration but introduces an improved, multi-cylindrical model that takes account of variation in the resistance of the stele, cortex and other tissues to oxygen diffusion. The other model, on convective gas flow, evaluates the importance of convection and diffusion in root aeration.

It seems, therefore, that land plants are adapted differently to overcome oxygen stress in the subterranean zone, i.e. by producing more root hairs (if the soil is dry) or producing special respiratory roots with or without root hairs. These root hairs grow out or upwardly towards the oxygen source (if flooded) or carry out anaerobic respiration or sometimes exhibit all these adaptations in combination to become more successful inhabitants of hypoxic or anaerobic environments. In typical hydrophytes, it may be much more efficient to store oxygen in air passages than producing root hairs, while in terrestrial plants, root hairs are the chief avenues for oxygen intake. The question arises as to why most of the land plants have not developed such a well-developed air passage system in their shoot and root like the hydrophytes. If prominent air passages had developed in the land plants, it would have probably reduced the whole mechanical strength of the standing plant body. The exposed shoot surfaces can, however, maintain normal respiration by utilizing the oxygen generated by photosynthesis and also oxygen diffused through stomates, but the underground parts suffer from more hypoxic condition compared to the shoot and hence in the absence of well developed internal air passage system the roots of land plants depend primarily on root hairs for their oxygen supply.

g) O_2-Dependent water and nutrient uptake by roots (active uptake of water and minerals)

There seems to be an intimate relationship between respiratory metabolism and absorption of water by root hairs. Higher the respiration rate, higher the absorption rate (Cailloux, 1972; Ahmed and Cailloux, 1971, 1972). When 127.5 μm of the tip of root hair is immersed in a capillary, it will absorb, whereas the same root hair immersed at a depth of 255 μm will excrete (Cailloux, 1950). Pollutants like flouride, Azide, CO_2 or any respiration inhibitory factors may depress root hair activity, kill root hairs and the plant may starve of water due to physiological dryness.

Levitt (1953, 1974), who proposed that absorption of water is purely a passive process, agreed with Rosene (1950) that oxygen is needed for maximum water absorption by roots in view of the need for active absorption of solutes to maintain root pressure. The present 'respiratory function hypothesis' of root hairs indirectly supports the active water uptake hypothesis at least in root hairs, as originally proposed by Rosene (1943) and Cailloux (1943, 1972), although this is not currently a well-accepted hypothesis. Many authors have observed that higher the respiration rate, higher the absorption rate (Cailloux, 1972; Ahmed and Cailloux, 1971, 1972).

h) How do aquatic plants respire? (evidences from hydrophytes)

Additional evidence is available from the comparative anatomy of terrestrial plant root systems and the root systems of hydrophytes. It is well established that most plants occupying wetland or aquatic habitats (hydrophytes and halophytes) lack root hairs in their submerged parts, but develop organized internal gas pathways that allow the movement of oxygen in the gas phase (Sculthorpe, 1985; Curran *et al.*, 1986). The generalized absence of root hairs in hydrophytes is ordinarily attributed to the submergence of a part or the entire plant body in water. Thus, hydrophytes did not evolve any special organs to facilitate water absorption. Submerged hydrophytes are generally devoid of a thick cuticle and stomata on the outer epidermal layer, however, stomata may develop when the epidermal layer is exposed to light and air.

The absence of root hairs as special respiratory cells in certain submerged hydrophytes can be attributed to their development of internal air passages to store oxygen and maintain normal respiration. Since root hairs are functional only in the presence of gaseous oxygen, they do not simply develop under water where they have no function to perform in the absence of gaseous oxygen, and hence root hairs are absent.

Higher plants rooted in water-logged soils and aquatic sediments have a well-developed system of air lacunae, permitting efficient intra-plant transport of gases, including oxygen (Sculthorpe, 1967; Armstrong, 1982). The lacunal volume often constitutes about 50% of the tissue volume in hydrophytes (Raven *et al.*, 1988; Hostrup and Wiegleb, 1991). Hydrophytes are adapted to release oxygen from photosynthesis into the internal air passages and, in turn, utilize it for their respiratory activity. High oxygen concentration (10-20% v:v) could be found in the gas phase of roots or rhizomes of intact plants well adapted to wet habitats or standing water. Removal of the aerial shoots, however, caused an almost complete exhaustion of oxygen within the submerged parts (Monk and Brandle, 1982). Whilst organs deprived of light generally have lysigenous lacunae, some roots and rhizomes possess the schizogenous spaces more typical of illuminated photosynthetic organs. The small schizogenous spaces of the young roots of *Callitriche* and *Myriophyllum* are later enlarged by the breakdown of the septa between them. The location of air lacunae is only species-specific. Further, roots of certain hydrophytes frequently release oxygen to the surrounding hydro-soil (Armstrong, 1979; Sand-Jensen *et al.*, 1982; Christensen *et al.*, 1994). The oxygen release per cm of root during illumination was about twice as high in the young root compared with the older root. The lower outward flux from the old root was ascribed to a reduction in root wall permeability with age.

In certain floating aquatic plants which lack root hairs, such as *Pistia stratiotis*, the leaves are densely covered by uniseriately multicellular trichomes on both the sides of leaves which are found to act as conduits of air into the internal air passage system. These trichomes are filled with air and when pressed, this air is pushed to the internal air cavities in the mesophyll (Bhaskar, unpublished data). Likewise, different aquatic plants seem to have developed different adaptive structures to enable oxygen diffusion into the submerged parts and store oxygen to be used as and when required.

It is said that diffusion of oxygen through water is 1/10,000 of its diffusion rate in air (Curran et al., 1986). According to Drew (1979), O_2 diffusion in pure water is about three million times slower than in air. Hence, recycling of oxygen from photosynthesis is an efficient modification in hydrophytes. Aerial foliage of hydrophytes is considered to be the main source of oxygen for the organs buried in the substrate (Samantarai, 1938; Conway, 1937; Laing, 1940; Coult and Jallance, 1951, 1958). Of all the vegetative organs in various species of hydrophytes, the leaves possess the highest oxygen concentration (10-19.6%). The evolution of oxygen during photosynthesis produces a gradient of concentration extending down through the internal atmosphere of the petioles and leaf bases to the underground organs. At the depths inhabited by underground stems and roots, submerged soils are generally quite devoid of O_2. Even in the overlying water, the O_2 concentration may frequently fall to 0.4% or less. The O_2 concentration in the rhizome and root of these various species has been found to be as low as 2 or 3% (Sculthorpe, 1985). The existence of linear gradient of O_2 concentration from the aerial to the underground parts supports the hypothesis that underground organs derive their O_2 supply from the aerial or floating foliage or from upper parts of plant body close to light or air. The rate of transport of O_2 will be influenced primarily by concentration gradient (i.e. by the relative rates of net photosynthetic O_2 production in the foliage and of respiratory O_2 consumption in the underground organs (Sculthorpe, 1985).

Guttenberg (1968 cf. Ellmore, 1981) compared the upward-growing roots (UGR) in *Ludwigia peploides* (Onagraceae) to the pneumatophores of the mangrove *Avicennia*, which connect underground gas spaces in the plant to the atmosphere. The studies of Ellmore (1981) gave firm evidence that gas can easily pass between stems and UGR. First, the novel O_2 and CO_2 content of gas released from UGR is identical to that which escapes from the stem. Second, most gas escaping from UGR must come from the stem because large amounts (up to 25 ml/hr) are released and the gas stream abruptly stops if the root is severed from the stem. O_2 and CO_2 composition

(% by volume) of gas released from UGR and shoots of *Ludwigia* during light and dark periods is given below.

Gas	Air	Roots		Shoots	
		Light	Dark	Light	Dark
O_2	21%	14%	7.1%	14%	7.1%
CO_2	0.03%	0.21%	2.13%	0.21%	2.13%

(cf. Ellmore, 1981)

Aeration may involve the release of CO_2 as well as the uptake of O_2. The latter is more significant because CO_2 toxicity in plants does not appear until its concentration nears 10%, a level never approached in *Ludwigia* (Ellmore, 1981).

Several investigators have analysed the gases in roots of plants growing in water or mud which is poor in oxygen.

According to Raskin (1985), mass flow of air to the submerged parts of the plant constitutes the major mechanism of aeration in partially-submerged rice plants. He proposed that the flow of air results from the reduction of pressure in the air conducting system of the plant caused by conservation of oxygen and solubilization of respiratory carbon dioxide in the surrounding water. This carbon dioxide solubilization theory and consequent mass flow of air from aerial parts to the submerged parts can function in light as well as darkness and may also operate in other semi-aquatic plants. This conflicts with the thermal transpiration and hygrometric pressurization theory as proposed by Dacey (1980) that requires heat from sunlight to initiate pressurization in the free space of the aerenchyma tissue. The theory of mass flow of air due to CO_2 solubilization might also function to a certain extent in the case of terrestrial plants and it is worth testing if this mechanism is operational in land plants, especially in flood-tolerant species.

In mangrove swamps, air roots called pneumatophores protrude upwards in great numbers in order to gather air into the root system which is submerged in the mud. A single tree may produce several thousand of these air roots, usually 20-30 cm high and 1-2 cm thick, soft and spongy, and studded with a number of little whitish lenticels or pores. In the mud, they connect to radially-running main roots, which are also soft and spongy and contain large amounts of air (Scholander *et al.*, 1955). When the lenticels are freed by the falling tide, the oxygen rises in the roots until they are covered by the next high tide. This is caused by the air being drawn in through the lenticels on the pneumatophores. When the pneumatophores are eliminated, the oxygen in the roots drops, reaching 1% or less in two days, proving that the pneumatophores in *Avicennia* are ventilating ports for the root system in the anaerobic mud. In *Rhizophora*, stilt roots end in a bunch of spongy, gas-filled roots embedded in the mud. There is direct gas connection between the mud roots and the lenticels on the stilt roots. When the lenticels were plugged by grease, the oxygen in the buried roots fell and reached 2% or less in two days. The high oxygen tension in the roots of *Rhizophora* is, therefore, maintained by means of a ventilation which takes place through the lenticels on the stilt roots (Scholander *et al.*, 1955).

Contrary to the notion that all hydrophytes lack root hairs, the roots of certain submerged hydrophytes are reported to develop abundant and very long root hairs (Shannon, 1953; Sculthorpe, 1985; Bagyaraj *et al.*, 1979; Clayton and Bagyaraj, 1984; Bhaskar and Kushalappa, unpubl.). Then the question arises, if the function of the root hairs is respiration, how to explain the respiratory role of aquatic root hairs submerged within water? Bhaskar and Kushalappa (unpubl.) have discovered that *Alternanthera sessilis*, which is a semi-aquatic or an emergent aquatic plant, produces dense and long root hairs only on the new long roots produced from the nodes just below the water level, while the nodal roots produced at deeper depths are without any root hairs. Further, root hairs in *A. sessilis* were produced only in fresh water with more dissolved O_2 than in polluted stagnant water with high BOD. This is a clear evidence to support that the hydrophytes produce root hairs mostly closer to O_2 source, i.e., air or at upper reaches of water with more dissolved O_2. This can be attributed to the linear gradient that exists with regard to dissolved

O_2 content from surface to the bottom of the water pool. If their main function was not respiratory, these root hairs would not have been produced nearer to air. But the only difference between the root hairs of terrestrial plants and hydrophytes would be that in the former, root hairs are adapted to grow in voids with humid air with gaseous O_2 or water droplets with dissolved O_2, while the latter, i.e. root hairs in hydrophytes are adapted to grow in a liquid form of water with higher concentrations of dissolved O_2.

Interestingly, the root hairs in aquatic plants seem to differ in their cell wall structure from that of root hairs in terrestrial plants (Emons and van Maaren, 1987). In the root hairs of most of the terrestrial plants, the outer layer has random micro-fibrils and the inner layer with axially-oriented micro-fibrils. The micro-fibrils in the outer layer form a loose network. But most aquatic plants have a helicoid cell wall texture in the secondary wall (inner layer), whereas in the tips of root hairs, only the primary walls (outer layer) with randomly–oriented micro-fibrils are present. There are an increasing number of helicoidally arranged lamellae presenting a counter clockwise mode of rotation that is species-specific in most aquatic plants. The presence of helicoidal or axially-oriented micro-fibrils in the inner wall layer seems to have some functional and adaptive significance in view of their development in liquid medium.

On the basis of the above findings, the author recognizes two kinds of root hairs, viz., aerial root hairs and aquatic root hairs, but in both the cases the chief function of the root hairs remains the same, i.e. intake of oxygen.

i) How do roots of terrestrial plants respire? (evidences from terrestrial plants)

Plants which have comparatively few root hairs may compensate by having an efficient internal air passage system. Tuberous or bulbous Liliaceous plants (*Gloriosa, Iphegenia*) have been shown to develop air passages in the basal inter-node which connect the underground bulb to the aerial atmosphere in the axil of the first aerial leaf. The air passage has an internal layer of epidermis which has stomata (Dayanandan *et al.*, 1986). These species require a sufficient oxygen supply in the bulb and root hairs alone may not be present in sufficient numbers.

Subterranean roots are quite comparable to submerged hydrophytes, in the sense that roots have a thin cuticle and no stomata, but may develop a thick cuticle as well as stomata when exposed to sunlight. Normally, terrestrial roots have thin cuticle (Dawes and Bowler, 1959) because the epidermal layer is only in contact with air through the soil pore space. Presence of cuticle on root hairs is debatable, although Dawes and Bowler (1959) have demonstrated it cytochemically in radish (*Raphanus*). However, none of the recent studies on cell wall texture of root hairs (Sasson *et al.*, 1985; Emons and van Maaren, 1987; Mort and Grover, 1988) has indicated the presence of a cuticle.

j) Stomata in root hair zone

Stomata are commonly present on most aerial parts. The stomata, if present on the roots, would facilitate the diffusion of oxygen into the roots, but due to the absence of light below-ground, stomata cannot open and close and hence roots generally lack stomata. However, occurrence of permanently open stomata in the root hair zone of seedlings has been reported (Tarkowska and Wacowska, 1988). Presence of stomata on seedling roots of *Helianthus annuus* was also reported by Tietz and Urbasch (1977) and Lefebvre (1985) in *Pisum sativum* and on the primary roots of *Ceratonia siliqua* by Christodoulakis and Psaras (1987).

Christodoulakis and Psaras (1987) reported that the stomata are randomly distributed throughout the surface of the root zones which possess mature vascular tissue, i.e. from the zone of root hairs to the transition zone. Tietz and Urbasch (1977) proposed that these stomata on roots are non-functional in nature. Based on the pattern of guard cell wall thickening, typical stomatal pore and substomatal chamber, Christodoulakis and Psaras (1987) proposed that these stomata are functional, evidently serving in some kind of fluid or perhaps gaseous exchange.

Tarkowska and Wacowska (1988) feel that the presence of the stomata in the root hair zone is of key importance. They observed that the stomata occurred in root hair zones in *Pisum, Ornithopus* and *Helianthus*. The stomata which were found in the root hair zone were almost always open, usually without chloroplasts and were insensitive to the action of ABA, unlike the stomata present in the upper part of hypocotyl or towards cotyledons. The stomata in the root hair zone were almost always without thickened cell walls at the border of the aperture (see Figs. 1-15 in Tarkowska and Wacowska, 1988). The stomata in the root hair zone, in addition to root hairs, may enable even greater diffusion of oxygen in highly active tissues where root hairs alone may not suffice. Once the primary root branches out and increases respiratory surface area the need for stomata may be alleviated, hence leading to their degeneration. Observations of Tarkowska and Wacowska (1988) on the occurrence of stomata in the root hair zone supports the present hypothesis that the root hair zone and especially the root hairs themselves are specialized for respiration.

5. ADDITIONAL EXPERIMENTAL EVIDENCES

Additional experimental evidences have been obtained supporting the author's theory that root hairs are specially adapted for intake of gaseous oxygen and to carry out respiratory function. It is further shown that root hairs are aerobic in occurrence and negatively hydrotropic. They are very sensitive under gaseous oxygen deficiency in soil environment. As per the indications of results obtained, it appears possible to increase the crop growth and yields by promoting root hair development through proper soil and water management.

The following experiments were conducted and the methods followed and the results obtained are presented below:

Expt. 1) Effect of flooding on root hair development

Tomato (*Lycopersicum esculentum*) crop was raised in the field with flat terrain and the soil was red sandy loam. The crop was irrigated every 4 days by the furrow method in one treatment, which created a more or less saturated condition and in the other, irrigation was restricted by 50%. Observations were made on the effect of an anaerobic condition created in the soil on the root hairs.

It was observed that in plots which were excessively irrigated, new whitish roots with dense root hairs grew out of the soil surface (**Fig. 9**), while the roots inside the soil with close contact with water were devoid of root hairs. In plots with restricted irrigation, tomato plants had roots confined inside the soil and also possessed root hairs.

Expt. 2) Effect of sand (with aeration) and broth solution (without aeration) on root hair formation and nodulation

Soybean (*Glycine max*) crop was raised in a leonard jar filled with sand and a wick was introduced to absorb water and nutrients from the broth solution kept in a jar. Observations were made on the presence or absence of nodules in roots in sand (obviously with more aeration) and those which had grown into the broth solution (without aeration).

The roots of soybean plants grown inside the sand medium had developed profuse root hairs and abundant root nodules, whereas the roots which had grown into the broth solution were completely devoid of root hairs and nodules.

It has been reported that in soybean grown under saturated soil culture, roots proliferated above the water table and also root nodules were concentrated above the water table (Guafa *et al.*, 1993). Normal irrigation is reported to enable better nodulation in *Trifolium subterraneum* L. by *Rhizobium trifolii* TA1 (Worrall and Roughley, 1976). The present finding possibly provides an answer to the question what promotes nodulation under a more aerated soil environment. The roots of soybean, when subjected to optimum soil aeration as in sand culture, promotes more root hair development and hence results in the increased chances of root hair infection and consequently, more number of root nodules as compared to roots immersed in liquid medium, which inhibits the development of root hairs.

Additional evidence is available from the study of Saur *et al.* (1998) who found that in case of a swamp forest species (*Pterocarpus officinalis*), majority of the root hairs (and incidentally nodules) were found above the water table, located on large aerial buttresses. Large trees were found to have modified their environment by accumulating litter between the buttresses, ensuring that a certain amount of soil above the water table. Consequently, the root hairs and the nodules were concentrated in a circle 5 metre in diameter around the oldest trees. A few nodules (5%) survived below the water table level, provided that healthy root hairs were present.

Why do N_2-fixing microorganisms infect through root hairs?

Invasion of nitrogen-fixing microorganisms and VAM takes place clearly through root hairs. This poses a question: why through root hairs? Nitrogen fixation by Rhizobia in the infected cells of legume nodules is directly dependent on a high rate of ATP synthesis through bacterial oxidative phosphorylation (Thumfort *et al.*, 1994). Therefore, in nitrogen-fixing legume nodules, the oxygen is required in large amounts for aerobic respiration. Nodules typically respire at about 4 times the rate of an equal biomass of roots (Layzell and Hunt, 1990). It has been shown by the author through several earlier illustrations that root hairs are the specialized cells through which, the oxygen diffuses from the soil atmosphere to the root and also that root hairs are the centres of ATP production in view of the densely-studded mitochondria with maximum cristae development (See Chapter 6). Root hairs develop where aeration is prevailing and obviously, oxygen is available. This provides a favourable environment for the growth and metabolism of the microorganisms. In legume root nodules, outer parts of nodule are respiratory and hence oxygen is required for the bacterial respiration and the consequent transport of ATP to N_2 fixing sites through the plasmodesmata. The fine structure of the non-infected (outer) region of a soybean root nodule is described by Sprent (1972). The cells are mainly vacuolated, with active cytoplasm. They are connected with each other as also cells of the infected region by numerous plasmodesmata. A network of air spaces runs throughout the nodule. Clearly, the nodule structure must facilitate an interchange of material, so that the nitrogen reaches the bacterioids freely and O_2 is used to manufacture ATP required for N_2 fixation. The intercellular space system which penetrates from the outside to the centre of the nodule would serve the purpose. Under O_2 deficient conditions, which might be less in soil, the O_2 concentration reaching the centre would be much less than that in the atmosphere. Indeed, as Bergersen (1962) showed, nodules are usually limited in the extent of their N_2 fixation by O_2 supply. Translocation of O_2 from shoot to the root seems unlikely. Thus, movement across the nodule appears the most important route for O_2 transport. The O_2 sensitivity of nodule activity argues for a positive role of the host cell. Further, the roles for leghaemoglobin have included O_2 binding among others. Without leghaemoglobin diffusion of O_2 through the dense nodule tissue, it would be completely inadequate to meet the ATP requirement. The properties of the haemoprotein permit the flux of O_2 to be maintained through the tissues, but the rate of consumption by the bacteria is too great to all the accumulation of a significant free O_2 concentration (Bergersen, 1971). Thus, the close association of nitrogen-fixing microorganisms with root hairs further supports the hypothesis that the root hairs are the main entry points for oxygen diffusion in roots.

However, nitrogenase, the bacterial enzyme that fixes N_2, is oxygen labile. Of late, a lot of focus has been laid on the aspect of maintaining an appropriate balance. Layzell and Hunt (1990) have found that a high rate of oxygen consumption and a cortical barrier to gas diffusion appears to work together to maintain the concentration of this element in the infected cell at a level that is not inhibitory to nitrogenase activity. Actinorhizal plants are reported to show a large range of anatomical and biochemical adaptations in order to balance the oxygen tension near nitrogenase (Huss-Danell, 1997). In symbioses with well-aerated nodule tissue in *Alnus*, the vesicles have a multilayered envelope composed mainly of lipids, bacteriohopanetetrol and their derivatives. This envelope is assumed to retard the diffusion of oxygen into the nitrogenase-containing vesicle. In symbioses like *Casuarina*, the infected plant cells themselves, rather than *Frankia*, appear to retard oxygen diffusion, and high concentrations of haemoglobin indicate an infected region with a low oxygen tension (Huss-Danell, 1997).

A low diffusion of oxygen in the poorly drained soils of the humid pasture, which remained wet throughout the year, was attributed to the abrupt decline in mycorrhizal colonization (Barnola and Montilla, 1997). They also showed that in well-drained soils in the shrub-rosette site during the dry season were less fertile and mycorrhizal colonization was the highest ($69.4 \pm 2.5\%$) compared to the more poorly drained pasture soils.

Expt. 3) Effect of submergence on root hairs

In order to prove that root hairs are negatively hydrotropic, two experiments were conducted. In one experiment, a drop of water was put on dense root hairs on the primary root of soybean seedlings grown inside a glass petridish with moist air and constantly observed under the binocular research microscope for 4 hours for any effect on root hairs. In the other experiment, seeds of finger millet (*Eleusine coracana*), paddy (*Oryza sativa*), sweet pea (*Pisum sativum*) and soybean (*Glycine max*) were sown on paper towels kept inside petridishes. In one treatment, the paper towel was flooded with water to submerge seeds while in the other, the paper towel was just moistened with water to create an aerobic environment for seeds. Observations were made at regular intervals on the development of root hairs and its effect on growth of seedlings.

In soybean and pea seedlings, the portion of root hair zone which was submerged in a drop of water resulted in complete disintegration of root hairs within few hours after submergence. Interestingly, the drop of water was also not absorbed by the root hair zone and remained for several hours (4 h). Further, no fresh root hairs were produced from this root hair zone damaged by submergence even after the water droplet dried. Root hair cells have been reported to maintain vacuolar acidity for at least 2 h during anoxia and ADP accumulated during anoxia (Brauer *et al.*, 1997).

In the other experiment, the seedlings of finger millet, sweet pea, maize and soybean grown on damp paper towel showed an enhanced presence of root hairs on the radicle above the paper towel and exhibited advanced growth when compared to seedlings grown in flooded condition which were completely devoid of root hairs (**Fig. 7**).

In an another experiment, seeds of maize (*Zea mays*) were sown inside the petridish using a wet blotter paper. After 3 days, seeds germinated, showing radicles with profuse root hairs exposed to the humid air inside the petridish. In one trial, hairy radicles were completely immersed in water for two days to observe the effect on root hairs. It was seen that under submergence, root hairs remained intact for 24 hours but obliterated after that. Meantime, a number of adventitious roots were produced—for which maize is well known for—among them those produced under submergence lacked root hairs or only a few sparsely-produced root hairs could be seen, while the roots produced above the level of water had developed dense root hairs. This experiment has indicated that the life span of root hairs after submergence is abruptly reduced but the time required for the degeneration of root hairs may be different in different species. Further, in maize, it is also shown that as a consequence of anaerobiosis due to submergence, it tends to produce more adventitious roots with internal gas cavities as a means to overcome the oxygen stress.

Expt. 4) Effect of oxygen deficiency on root hairs

In order to find out the effect of oxygen deficiency on root hairs, seedlings of *Eucalyptus* (*E. teretecornis* or locally called *E. hybrid*) were grown in plastic containers and watered. For one set of seedlings, the container with soil and root was enclosed in an another hole-proof empty plastic bag and its mouth was tightly closed with wax to make it airtight. This was done with the idea that the roots after exhausting oxygen inside the bag will be subjected to an oxygen deficient or anoxic environment. The other set of seedlings were kept outside with their ball of earth exposed to air. After a month, roots of the seedlings subjected to these treatments were examined for the presence or absence of root hairs.

It was observed that the roots of eucalyptus seedlings which were subjected to oxygen deficiency were completely devoid of root hairs, unlike the roots of seedlings which were kept exposed to air.

Expt. 5) Effect of soil type and restricted irrigation on root hair development

Teak (*Tectona grandis*) seedlings were grown in two kinds of soils: a heavy clayey soil and a fine sandy soil. Seedlings were watered once in 4 days in case of both treatments but in case of sand culture, the quantity of water was restricted by 50%. After about a month, the roots were subjected to observation for the nature of root hairs.

It was observed that the teak seedlings grown in heavy soils were devoid of root hairs as well as ectomycorrhizal association, whereas seedlings grown in sand with restricted watering had distinct root hairs and had also associated with ectomycorrhizal fungi. Incidentally, the seedlings which were grown in sand produced a new flush of leaves and grew better than those grown in heavy soils.

Expt. 6) Effect of oxygen enrichment in soil (through restricted irrigation) on growth and yield of paddy

Two experiments were conducted to study the effect of oxygen enrichment in soil through an increasing porosity in the medium on growth and yield of paddy. Oxygen enrichment was achieved by changing the medium from normal soil to leaf litter (*Acacia nilotica*) and restricting irrigation to soil. The former was conducted in the greenhouse and the latter in the field.

In the first experiment, a pot experiment was conducted in a field lab at Bangalore to study the soil medium on the growth of paddy. The media used were the humus soil from *Acacia nilotica* plantation (T_2) and leaf litter (without soil of course) (T_3) collected form *A. nilotica* plantation of 13 year age. Soil from an adjoining open field was used as control (T_1). Pots measuring 12" × 9" were filled uniformly. Each treatment had four replications. These pots were regularly watered to obtain compactness. Later, a known number of presoaked paddy seeds (25 Nos.) of uniform size were sown in the pots at a dept of 2.5 cm. Pots were regularly watered but water stagnation was not allowed in all the treatments. Weekly observations were made on the growth rate of paddy. The results are given in the **Table 2 (Fig. 10a)**. As observed, maximum growth parameters were recorded in paddy grown in litter (T_3) (obviously which had no available nutrients) followed by treatment with humus soil from *A. nilotica* plantations (T_2) and soil from adjoining area (control—T_1). Increased growth of paddy in a pure litter medium is due to an increased porosity and concurrent oxygen enrichment to root hairs rather than the availability of more water or nutrients.

The petridish experiment on the effect of water quantity on the development of root hairs in paddy (see Expt. 4) has shown that higher water regimes (flooding) suppress the development of root hairs and lesser water regimes (i.e. just moistening the blotters) promote dense root hair development. Thus, the porous soils promote oxygen enrichment and thereby development of root hairs which chiefly enhance the root hair respiration rather than water and mineral absorption.

Another experiment conducted on paddy in the field has confirmed that grain yields could be enhanced by at least 13% in paddy by such agronomic practices which promote root hair development and concomitantly oxygen enrichment. The treatments applied were irrigation to paddy (var. 'Ponni') once in 3 days as per the standard practice (T_1) and irrigation once in 6 days (T_2). The former treatment resulted in 1.7 tonnes per ha whereas the latter treatment yielded 1.93 tonnes grain per ha., i.e. an increased yield by 13.5% (**Table 3** and **Fig. 10b**).

The earlier notions that plant water deficits reduce the crop yields is untrue and mild water deficits can, in fact, actually enhance yields (Hearn and Constable, 1981; Rawson and Turner, 1982) through the maintenance of optimum soil moisture and oxygen. Rawson and Turner (1982) showed that in sunflower, withholding irrigation water for the first 45 days resulted in the production of larger leaves (up to 60% greater) in the upper canopy compared to the frequently-irrigated plants. Seed yield was also linearly correlated with maximum leaf area per plant.

There are several instances of increased yields due to drip or sprinkler or alternate furrow irrigation methods as compared to flood or surface irrigation methods (Hearn and Constable, 1981; Rawson and Turner, 1982; Turner, 1990; Graterol *et al.*, 1993) (**Table 1**). Several crops are known to yield higher yields (20 to 50%) when soil moisture deficiency was maintained at 50% of the field capacity (Saksena, 1992; Shanmugam, 1992). Under restricted soil moisture, grain yield is reported to be higher (15-18% in chickpea) (Nanda and Saini, 1992) and crop yields increased often up to 180% as in case of water melon in Jodhpur, India (Saksena, 1992) than the crop raised under high soil moisture. In the present study also, it was found that soil aeration by restricting irrigation enhanced the growth rate of finger millet, soybean and teak plants.

The increase in crop yields due to restricted irrigation as reported by earlier workers has been most commonly interpreted as due to continuous and optimum supply of water to roots without causing water stress during the growing period of crop plants. But higher yields due to restricted irrigation methods, such as drip irrigation method, can be actually attributed, in addition to maintaining optimum soil moisture, to creation of an adequate soil aeration, better root hair development, more O_2 supply to root hairs to enable root hair respiration and release of energy required for nutrient uptake and plant growth.

All these findings are a clear indication that saturated soils or surface irrigated soils would dampen and damage root hairs or prevent root hair development because these conditions are similar to the anaerobic environment which are unfavourable for root hair development, thereby affecting root hair respiration, root nodulation (in case of N_2-fixing species) nutrient metabolism and thereby reducing plant growth. By surface irrigation, roots will suffer from respiratory stress until the new roots are produced towards less moist

Table 1: Increased crop yield due to drip and sprinkler irrigation over conventional (surface or flood) irrigation methods

Crop	Average crop yield/ha			% increase	Author	Locality
	Conventional	Sprinkler	Drip			
Sugarcane	65 MT		100 MT	35	Saksena (1992)	Rahuri (Mah.)
Banana	14.50 MT		29 MT	50		"
Pomegranate	75,000 fruits		1,09,000	45		"
Mosambi (sweet lime)	1,00,000 fruits		1,50,000 fts	50		"
Grapes	26,400 kg		32,500 kg	23		"
Groundnut	2675 kg		3200 kg	19		"
Tomato	32,000 kg		48,000 kg	50		"
Cotton	2330 kg		29.50 kg	27		"
Tomato	164 kg		171.86 kg	5		"
Brinjal	280 kg		320 kg	14		"
Bitter gourd	154.34 kg		214.71 kg	39		"
Ridge gourd	171.30 kg		200 kg	16		"
Onion	93.0 qu.		112 qu.	20		
Potato	235.7 qu.		344.2 qu.	46		Dapoli (Mah.)
Okra	157.7 qu.		177.2 qu.	12		
Sugarbeet	NA		NA	30		Maharashtra
Pomegranate	NA		NA	30		Maharashtra
Guava	NA		NA	25		Maharashtra
Custard apple	NA		NA	20		Maharashtra
Groundnut	1713 kg	1954 kg	2842 kg	65		Vadodara
Tomato	6187 kg		8872 kg	43		Jodhpur
Okra	10,000 kg		11,310 kg	13		Jodhpur
Beetroot	571 kg		887 kg	55		Jodhpur
Raddish	1045 kg		1186 kg	13		Jodhpur
Brinjal	12,400 kg		12,300 kg	– 0.80		Jodhpur
Chillies	4233 kg		4080 kg	– 3.60		Jodhpur
Sweet potato	4244 kg		5888 kg	38.70		Jodhpur
Sugarcane	87,000 kg		75,000 kg	– 13.8		Jodhpur
Bottle gourd	38.01 t		55.79 t	46.7		Jodhpur
Ridge gourd	10.74 t		12.03 t	12		Jodhpur
Round gourd	29.47 t		36.60 t	24		Jodhpur
Watermelon	29.47 t		82.33 t	179.36		Jodhpur
Potato	20.18 t		33.58 t	66.40		Jodhpur
Groundnut	2380 kg	2610 kg		9	Shanmugam (1992)	Rahuri (Mah.)
Chillies	1890 kg	2420 kg		28		
Chickpea				15-18	Nanda & Saini (1992)	

Table 2. Influence of soil and litter of Acacia nilotica plantation on growth of paddy

Treatment	Germination (%)	Shoot height (cm) Days after sowing			Number of leaves/plant Days after sowing			Leaf area (cm^2/plant) Days after sowing		Number of tillers/plants at 46 DAS*
		18	32	46	18	32	46	32	46	
Soil under Acacia nilotica	21.00	3.99	4.35	6.69	3.08	3.69	3.78	6.59	8.25	0.14
Leaf litter of Acacia nilotica	20.00	4.53	6.82	10.70	3.21	3.81	8.63	10.00	27.33	1.38
Control (soil from adjoining open field)	20.00	2.80	2.84	4.40	2.19	3.02	3.28	3.62	4.88	0.00
F-test	NS	**	**	**	**	**	**	**	**	**
S.Em. ±	1.30	0.13	0.94	0.43	0.06	0.12	0.74	0.77	2.86	0.23
C.D. (P=0.05)	-	0.43	3.00	1.36	0.20	0.39	2.38	2.48	9.14	0.75

* DAS = Days after sowing
** Significant both at P = 0.01 and P = 0.05
NS = Non-significant

Table 3: Effect of reduced levels of irrigation on yield of paddy

Treatment	Straw yield /ha. (tonnes)	Grain yield /ha. (tonnes)
T_1 = Watering once in 3 (control) days	1.78 (+7.9%)	1.7 (-13.5%)
T_2 = Watering once in 6 days	1.65 1.93	1.93 (+13.5%)

but relatively aerobic soil layers or voids. In such cases, relatively more current assimilates are apportioned to below-ground parts, resulting in the proliferation of roots. This kind of additional root growth would consume extra energy thereby affecting normal plant growth.

Oxygen deficiency is not limited to plants growing in flooded conditions; it also occurs in non-wetland soils. Many soils are poorly aerated because of compaction, surface crust formation (as in red soils) or continuous or excessive irrigation or rainfall. A low diffusion of oxygen in the poorly drained soils of the humid pasture, which remained wet throughout the year, may explain the abrupt decline in mycorrhizal colonization (Barnola and Montilla, 1997). Hence, in order to promote the respiratory dependent functions of roots, the chief agronomic practices including soil and water management need to be oriented towards promoting normal root hair growth, root hair respiration, normal symbiotic associations and healthy plant growth.

Therefore, to maintain proper soil aeration and normal functioning of root and root hairs, restricted irrigation methods such as furrow or alternate furrow irrigation, sprinkler and drip irrigation methods are primarily essential to ensure a more efficient water use and increased crop yields. Perhaps, future research work in crop production is required to be oriented towards providing a better soil environment for promoting normal root hair development, oxygen diffusion through root hairs and increase crop growth and yields.

This is one natural way of organic farming, aiming at increasing crop yields without any additional input costs in the form of fertilizers or irrigation. Incidentally, water is also saved and the water thus saved could be used for extending lands under irrigation and thereby, increase crop productivity.

CHAPTER 7

Implications of the Respiratory Role of Root Hairs in Agriculture and Forestry

How much oxygen is required for root (hair) respiration? How much CO_2 is released through root (hair) respiration? How are the roots adapted to overcome soil anaerobiosis? How do crops respond to flooding and restricted irrigation? Can root hair management enhance crop yields? These are some of the questions which need to be probed in view of the respiratory role of root hairs.

1. CONTRIBUTION OF ROOT (HAIRS) TO CO_2 EVOLUTION

Of all factors, root respiration is the most sensitive aspect of plant activity in regard to soil aeration. Since growth of roots and uptake of water and nutrients are dependent upon energy which is supplied by respiration, it may be assumed that reduction in respiratory activity is the first step in the growth-limiting effects of insufficient aeration.

Root respiration may account for more than half of the CO_2 evolution from forest soils. In deciduous forests in Japan, the proportion of CO_2 evolution from the A_0 layer, mineral soil, and roots were estimated to be 14 to 16, 44 to 46, and 20 to 40%, respectively (Katagiri, 1988). Root respiration accounted for 51% of soil CO_2 evolution in a 9-year-old *Pinus elliottii* stand in Florida and 62% in a 29-year old plantation.

It is now a well-known fact that most root respiration occurs in the fine roots (Kozlowski, 1992) and precisely, as per the present new hypothesis, the root hairs. The fine roots accounted for more than 95% of the root respiration of *Pinus sylvestris* and *Betula* stands (Mamaev, 1984). The sink strength of mycorrhizal roots (which have replaced root hairs and root hair zone in roots) is reported to be generally very high, largely because of the high respiration rates of the fungi in association with their large cytoplasmic volume and number of mitochondria, high soluble protein content, and activity of several enzyme systems (Barnard and Jorgensen, 1977; Smith and Gianinazzi-Pearson, 1988; Kozlowski, 1992). It has been estimated that mycorrhizal fungi (which may comprise only about 4% of the root system biomass) account for approximately 25% of the CO_2 respired by mycorrhizal roots (Harley, 1973; Phillipson *et al.*, 1975).

2. CO_2-O_2 RELATIONSHIP AND THEIR EFFECT ON ROOT HAIR RESPIRATION, PLANT GROWTH AND YIELDS

Good gas exchange between soil and the ambient atmosphere is very important to maintain an appropriate soil atmosphere with regard to O_2/CO_2 balance. Harris and van Bavel (1957b) showed that a concentration of 10% O_2 decreased respiration from that in air and after 15 days, roots receiving 15% O_2 and 5% CO_2 had a respiration rate which was 68% of the rate for roots which were exposed to air.

Low rates of gaseous diffusion in heavy, over wetter soils resulted in excessive CO_2 accumulation and lower O_2 in the soil profile. Within a soil profile, the concentrations of CO_2 was generally observed to be lowest at a depth of 10 cm, where it was sometimes less than 0.1%. With increase in depth, the percentage of CO_2 increased and becomes maximal at the 20 to 30 cm depth where it seasonally peaked at more than 5%. Below 30 cm in the profile, CO_2 often decreased but usually remained higher than in the surface layer (Buyanovsky and Wagner, 1983).

Under wheat, periods of highest CO_2 (6-8%) corresponded to times of intensive decomposition of plant residue within the soil profile.

Under corn and soybean, highest CO_2 concentration was related to the periods of intensive plant growth. Soil moisture and soil temperature combined, were found to be responsible for just more than 50% of CO_2 fluctuations. The influence of water tension on CO_2 in the soil atmosphere was more significant (Buyanovsky and Wagner, 1983).

Biological processes in the soil lead to the accumulation of CO_2 and a concomitant decrease in O_2. The rate of CO_2 accumulation depends both upon the soil biological activity and physical characteristics such as porosity, structure, aggregate size, moisture content and temperature. Root respiration of tobacco was reduced to 68% of the regular rate when soil air contained 5% CO_2. A concentration of CO_2 greater than the O_2 concentration was shown to be harmful to normal growth of plants (Harris and van Bavel, 1957a and 1957b).

Perhaps the least studied edaphic characteristic is the gas phase in non-water logged soils. Concentrations of CO_2 at levels that effectively modify or suppress root growth exist within the plough layer, with the highest concentrations at the plough depth (Buyanovsky and Wagner, 1983; Ycas and Zobel, 1983). Very little is known of the dynamic effects of these CO_2 levels except for some relatively recent data (Arteca *et al.*, 1979; Arteca and Dong, 1981), which suggest that the CO_2 concentrations at the root surface can and do modify even the rates of photosynthesis.

Due to compaction of the forest floor by timber operations, a reduction of soil porosity by more than 60% and reduced infiltration of water has been reported (Chauvel *et al.*, 1991; Nirmal Ram *et al.*, 1993). The need of O_2 in well-aerated soils is emphasized for certain trees like teak. Abundant fine roots form on young teak plants in the uppermost soil layer during the monsoon period. These largely die off in the dry season when they are replaced by new ones developed in the deeper layers—provided the soil aeration is adequate (White, 1991). Root hairs can be induced when seedlings are cultured in coarse sand medium. In teak (*Tectona grandis*), seedlings cultured in clayey soils do not develop root hairs, but when transferred to sand culture, the root hairs developed almost throughout the length of fine primary roots and even they had better mycorrhizal infection (author's own observation).

The effects of water logging (thereby causing anaerobic root environment) on crop may be many. In cowpea, seed yields reduced by 50%, reduced nodule activity occurred, leaf dry weight and leaf area reduced and lower leaves abscised.

Flooding or water logging of the clay soils can reduce crop yields and profitability. Higher soil moisture under zero tillage can reduce production in the wet condition (Lindsay *et al.*, 1983; Simpson and Gumbs, 1992). Cowpea production was more sensitive to excessive moisture than maize. Bedding systems of field lay out in general improve the soil drainage and crop performance compared with flat planting on heavy soils. In heavy soils, a rotation of maize in the major wet season and cowpea in the minor wet season are therefore recommended (Simpson and Gumbs, 1992).

3. PLANT RESPONSES TO SOIL ANAEROBIOSIS

Oxygen deprivation is not limited to plants growing in flooded conditions; it also occurs in non-wetland soils. Many soils are poorly aerated because of compaction, surface crust formation (as in certain red soils) or continuous irrigation. Hence, roots may respire anaerobically, with oxygen limitation resulting in a greatly-reduced uptake of major mineral nutrients, in growth reduction, injury, and often death of plants (Kozlowski, 1982, 1984a, 1984b, 1992). Insufficient oxygen reduces the permeability of roots to water and accumulation of toxins (Drew, 1979; Bradford and Yang, 1981). Roots would grow rapidly under aerobic environment by making use of little carbohydrate and produce little ethanol and CO_2, while under anaerobic conditions, they would grow slower but use more carbohydrate and produce more CO_2 and ethanol ('Pasteur effect'). Increased anaerobic condition, as for example due to flooding, causes decreased carbohydrate reserves. The anoxia inhibits germination. The hypocotyl elongates faster in hypoxia than in air, whereas root elongation is inhibited when seedlings are deprived of O_2 (Gay *et al.*, 1991).

In past years, flood tolerance or tolerance to an oxygen-deprived environment has been held to depend either on metabolic adaptations or on morphological and anatomical changes that improve aeration. It is pointed out that successful strategies for survival in the majority of wetland plants involves both types of adaptation. The formation of the aerenchyma that permits the internal flow or diffusion of gases is itself believed to be caused by biochemical changes (Jackson et al., 1990).

In the absence of O_2, root hairs do not develop and plants tend to overcome O_2 limitation, or in other words, soil anaerobiosis, in various ways. Plant species are adapted differently to soil anaerobiosis by: (1) morphological adaptations by developing aerial roots or adventitious surface roots which assist in gas exchange of submerged roots (Hook, 1984); (2) production of aerenchyma tissues in submerged roots (Coutts and Philipson, 1978; Hook, 1984); (3) formation of hypertrophied lenticels on submerged stems and roots and basipetal movement of O_2 to the roots through aerenchyma; and (4) biochemical or physiological adaptations.

a) Morphological adaptations

In many flood-tolerant plants, the death of roots is compensated for by regeneration of new roots which are more succulent and fibrous than the original roots (Hook et al., 1971). The original root system does not function normally because of low O_2 tension (Hosner and Boyce, 1962; Gill, 1970). In order of declining tolerance to flooding, *Eucalyptus grandis*, *E. robusta* and *E. saligna* were correlated with variation in capacity to form adventitious roots (Clemens et al., 1978). *Melaleuca quinquenervia*, a flood-tolerant species produced new secondary roots from the upper portion of the tap root, lateral roots and submerged portions of stem which formed upwardly-growing fine roots that were largely confined to the surface soil layer and the water above the soil layer. Production of such roots appears to be an important adaptation to flooding (Sena Gomes and Kozlowski, 1980b). *Eucalyptus camaldulensis* produced many thick roots from the top of the tap root within 10-15 days after flooding, whereas *E.*

globulus did not (Sena Gomes and Kozlowski, 1980c). Several lines of evidence show that the amount of anaerobic respiration in roots of flooded plants is supplemented by oxygen transported from tissues above the water line. The deeper penetration of water logged soil and greater flood tolerance of *Pinus contorta* were attributed to its more efficient oxygen transport (Philipson and Coutts, 1980).

Flooding reduced respiratory capacity of the roots of the flood intolerant species (*Acer saccharum*) much more than the flood-tolerant species, *Acer rubrum* and *Taxodium disticum*. The respiration capacity of *Acer saccharum* declined rapidly during the first 8 days of flooding (Carpenter and Mitchell, 1980).

In a very wet soil, a widely spreading superficial root system would avoid excessive moisture, while in a soil with a more limited moisture supply, a deeply penetrating root system will best meet both water and oxygen needs. On a very wet soil, only shallow roots find enough air to continue growing. When maize roots extending downwards to moist air encountered a water surface, they changed orientation and grew horizontally or upwards (Molisch, 1985). This was considered to be evidence of 'aero-tropism' or response of root to gas concentration gradients. Guhmann (1924) grew tomatoes and sunflowers in pots which were flooded, flooded-and-aerated, and well drained. It was observed that the root system in the flooded pots were horizontally oriented but that those developed in the well drained and flooded-and-aerated pots grew vertically. It was concluded that O_2 deficiency induces more horizontal growth.

Many old roots of corn and sunflower die under full-flooded treatment but numerous adventitious roots develop in these conditions. Many new roots originate from the stems above the water level, some roots floated in water and did not go into the soil. In all cases, such roots were the shortest, straight, large in diameter and had root hairs (Yu et al., 1969).

The present author has found that root hairs are also 'aero-tropic' and are very sensitive to direct wetting, and once a part of root hair zone loses hairs, hairs never developed in the same site but could only be produced on the newly-produced primary roots formed in an aerobic soil environment.

The adaptation of the root morphology to prevailing soil conditions has been emphasized as an important feature in the uptake of water and nutrients from the soil. Under natural conditions in soils, the O_2 supply to the roots is controlled by the O_2 diffusion through the soil, and the rate of diffusion, depending largely on the soil water content. Root elongation has been found to be especially sensitive to aeration, O_2 being the essential and morphogenetically-effective part of the air (Erickson, 1946; Leonard and Pinckard, 1946). These observations indicate an increased number of laterals in non-aerated water cultures (Coult and Vallance, 1958), whereas other reports describe decreasing root branching with decreasing O_2 level (Ranson and Parija, 1955). Geisler (1965) observed that in pea, root elongation and root branching were strongly affected by the air supply, elongation being favoured by increasing aeration and branching activity by decreasing aeration.

Flooding often induces proliferation of lenticels on flooded stems and roots (Sena Gomes and Kozlowski, 1980a; Bhaskar and Srinivasa Rao, unpublished work). In some species (e.g. *Salix alba*), lenticels assist in aeration and also serve as openings through which ethanol, acetaldehyde and ethylene are released (Chirkova and Gutman, 1972).

b) Anatomical adaptations

Certain mesophytes are adapted to root anaerobiosis by developing an internal gas path system or aerenchyma. Following root anaerobiosis, wheat roots show an enlarged intercellular system, together with other morphological-anatomical changes (Trought and Drew, 1980; Wiedenroth and Erdmann, 1985; Erdmann and Wiedenroth, 1986). Wheat seedlings survive a limited period of root anaerobiosis quite well and show no alterations in the respiration activity of the whole root system (Erdmann *et al.*, 1986). On the other hand, these roots quickly exhaust the oxygen from a closed chamber.

Similar conclusions were drawn for the effectiveness of internal oxygen transport in maize. A larger ATP/ADP ratio (as a consequence of an internal oxygen transport) was detected in aerenchymatous maize roots compared with normal aerated ones (Saglio et al., 1983; Drew et al., 1985). Oxygen transport from the shoot may be important for the respiration only in the upper parts of the roots. Oxygen diffusion to the toot tip was not sufficient. Hence, growth was arrested (Erdmann and Wiedenroth, 1988). Therefore, the importance of internal transport in mesophytes like wheat or maize should not be over-estimated compared with helophytes, where the oxygen demand of the subterranean parts is mainly met by fast oxygen diffusion from the shoot (Steinmann and Brandle, 1981; Studer and Brandle, 1984). Intact roots deprived of an external source of oxygen stop growing and may even die.

The survival of growing roots following oxygen deprivation in the rhizosphere is related to their ability to secure a certain minimum influx of oxygen for respiration. Their immediate source of this oxygen is the intercellular space of the root, particularly in the cortex. In turn, a supply of oxygen to the intercellular space relies upon diffusion from the shoot system through inter-connected gas-filled aerenchyma (Armstrong et al., 1991). In the absence of an external oxygen source, living roots can consume all, or almost all, the oxygen dissolved in cellular water, in intercellular spaces and in the immediate environment.

Plants (mostly bog plants) having high porosity roots (e.g. 30%) have less dependence on soil aeration and may have more tolerance to poor soil aeration conditions and such plants would be well suited to flood irrigation and hydroponics systems (Luxmoore and Stolzy, 1972). Changes in root porosity may be the inherited characteristics of plants that could be adapted for root tolerance to excess water in the soil (Yu, et al., 1969).

Recent studies of Wiedenroth and Jackson (1993) have demonstrated that the internal reserves of oxygen are unable to sustain normal rates of respiration for more than a few minutes. Over the linear phase of consumption, oxygen was depleted at a rate of approximately 1.5 kPa min^{-1}. Three pools of oxygen available to the roots for respiration under soil are: (a) intercellular gas space (approximately 6% v/v) in wheat; (Stolzy, 1972), (b) oxygen dissolved in cell water (approximately 92% of the root weight is water; and (c) oxygen dissolved in the 0.1 mm-thick film of water that covers the root.

c) Biochemical and physiological adaptations

Under conditions of O_2 deficiency, carbohydrates may not be completely oxidized to CO_2 and water. Intermediate compounds then accumulate, and the amount of energy released is inadequate to support plant growth (Kozlowski *et al.*, 1991; Kramer and Kozlowski, 1979). Under anaerobic soil conditions, the permeability of root hairs is often altered, resulting in loss of ions by leaching (Rosen and Carlson, 1984).

According to one theory (Crawford's metabolic theory), under anaerobic soil conditions, species might try to adapt by: (1) reducing ethanol production; (2) venting of ethanol to the external medium; (3) transport of ethanol from tissues in poorly aerated regions to well aerated regions where it is metabolized; and (4) some tolerant species also delay or avoid accumulation of ethanol by diverting glycolytic intermediates to alternate end products such as, lactate, malate, succinate, r-aminobutyrate, and alanine. Ethanol produced in hypoxic roots may enter the transpiration stream and be carried up to the stem, where it could diffuse into the cambium. Thus, ethanol in stem cambium could be of root origin (MacDonald and Kimmeaer, 1991).

A large number of changes in soil and plants due to water-logging have been reported. The most common changes in soils are a lower rate of O_2 diffusion, induction of ethylene formation, and lowering of Fe and Mn. It is evident that plants undergo various physiological changes to cope with anaerobic conditions. One of the important physiological adaptations is that they possess protective mechanisms against higher concentrations of Fe and Mn in flooded soils (Talbot et al., 1987). For example, plants under flooded condition, leak O_2 from their root surface and oxidize the reduced form of Fe and Mn before they reach the vascular system (Jensen et al., 1967; Armstrong, 1979; Talbot et al., 1987). Under anoxia, root hair cells are reported to maintain vacuolar acidity for 2 h and accumulate more ADP during anoxia (Brauer et al., 1997).

4. ROOT HAIR MANAGEMENT FOR HIGHER YIELDS

Crops can suffer from excessive soil moisture as well as from its deficiency. One of the characteristics of water logged or saturated soils is a deficiency of oxygen required for root respiration. Letey et al., (1962) showed that reduced oxygen due to water logging due to heavy precipitation or excessive irrigation practices is very detrimental to crops until a good root system is developed. Since root growth is stopped by very low oxygen content, it is reasonable that a plant without a good root system would be more severely damaged than one which has already established a root system. Further, Letey et al. (1962) showed that less water was used by plants under low oxygen. This behaviour tends to make the situation worse. The only way to improve the oxygen supply in a water-logged soil is to remove some of the water. However, a plant growing under a water-logged condition transpires less, produces a poor root system and therefore, becomes ineffective in improving its own situation. This points out to the need for adequate drainage either by natural or artificial means or fallow wise irrigation practices.

The benefits of drainage on plant growth are clearly linked with those of O_2 supply. The benefits of drainage, therefore, will come from ensuring that the longer pore spaces are free of water for as much of the cropping season as possible, so that there is a normal root hair development and root respiration. The soil has a relatively small reserve of O_2 if it is suddenly cut off from the atmosphere by crusting or surface flooding or compaction. Thus it is not surprising that crop plants like potatoes die off quickly with no oxygen at the roots (Hawkins, 1962).

When crop fields are irrigated, root hairs after getting drenched in water degenerate within a day. The plant tries to produce new primary roots towards soil layers relatively free from excessive moisture and thereby enable development of root hairs. In other words, by completely drenching the root system as caused usually by surface irrigation method, nearly 75 to 90% of the surface area which conduits oxygen to roots is blocked and to that extent both maintenance and growth respiration are affected. In addition, the plants are forced to spend extra energy to produce new primary roots each time they were flood irrigated. Thereby, much of the energy used in shoot growth would be diverted towards new root formation. Hence drip irrigation or irrigation through implanted pipes or pitchers would give best results. Not only this should help in conserving water but also promote better soil aeration together with better aboveground biomass production.

There are clear indications that saturated soils or surface irrigated soils would dampen and damage root hairs or prevent root hair development because these conditions are similar to anaerobic environment which are unfavourable for root hair development, thereby affecting root hair respiration, root nodulation (in case of N_2-fixing species), nutrient metabolism and thus reducing plant growth. By surface irrigation, roots will suffer from respiratory stress until new roots are produced towards less moist but relatively aerobic soil layers or voids. In such cases, relatively more current assimilates are apportioned to below-ground parts, resulting in the proliferation of roots. This kind of additional root growth would consume extra energy, affecting normal plant growth.

Oxygen deficiency due to soil compaction has been reported to decrease overall plant growth in *Phaseolus vulgaris* by causing changes in root hair morphology, rates of cytoplasmic streaming which facilitate nutrient and water uptake (Alessa *et al.*, 2000). In view of the present theory of oxygen intake through root hairs, soil compaction is believed to create an oxygen-deficient soil environment, which will lead to further root physiological implications. Hence, in order to promote the respiratory-dependent functions of roots, the chief agronomic practices including soil and water management need to be oriented towards promoting normal root hair growth, root hair respiration and healthy plant growth.

The earlier notions that plant water deficits reduce crop yields is untrue and mild water deficits can, in fact, actually enhance yields (Hearn and Constable, 1981; Rawson and Turner, 1982) through maintenance of optimum soil moisture and oxygen. Rawson and Turner (1982) showed that in case of sunflower, withholding irrigation water for the first 45 days resulted in the production of larger leaves (up to 60% greater) in the upper canopy compared to the frequently-irrigated plants. Seed yield was also linearly correlated with maximum leaf area per plant.

As already mentioned there are several instances of increased yields due to drip or sprinkler or alternate furrow irrigation methods as compared to flood or surface irrigation methods (Hearn and Constable, 1981; Rawson and Turner, 1982; Turner, 1990; Graterol *et al.*, 1993). Several crops are known to yield higher yields (20 to 50%) when soil moisture deficiency was maintained at 50% of the field capacity (**Table 1**) (Saksena, 1992; Shanmugam, 1992). Under restricted soil moisture, grain yield is reported to be higher (15-18% in chickpea) (Nanda and Saini, 1992) and crop yields increased often up to 180% as in case of water melon in Jodhpur, India (Saksena, 1992) than the crop raised under high soil moisture. In the present study also, it was found that soil aeration by restricting irrigation enhanced the growth rate of paddy, finger millet, soybean and teak plants and in case of paddy, there was an increase in grain yield by 13%.

The increase in crop yields due to restricted irrigation, as reported by earlier workers, has been most commonly interpreted as due to continuous and optimum supply of water to roots without causing water stress during the growing period of crop plants. But higher yields due to restricted irrigation methods, such as drip irrigation method, can be actually attributed, in addition to maintaining optimum soil moisture, to creation of adequate soil aeration, better root hair development, more O_2 supply to root hairs to enable root hair respiration and release of energy required for nutrient uptake.

Therefore, in view of these findings, for maintaining proper soil aeration and normal functioning of root and root hairs, restricted irrigation methods such as furrow or alternate furrow irrigation, sprinkler and drip irrigation methods are primarily essential in order to ensure more efficient water use and increased crop yields. It could also be possible to induce the development of more number and length of root hairs through microbial and chemical amendments. For example, inoculation with *Penicillium bilaii* in field pea is reported to increase the proportion of root containing root hairs by 22% and a 33% increase in the mean root hair length (Gulden and Vessey, 2000). Perhaps, future research work in crop production is required to be oriented towards providing a better soil environment to promote normal root hair development, oxygen diffusion through root hairs and increase crop growth and yields.

It is believed that in case of tropical wet evergreen forests occurring in heavy rainfall zones with almost continuous rain occurring from May through September, it would drench the roots, consequently plants in this forest may cease to produce root hairs. Roots would suffer from oxygen deficiency. Hence, the rate of growth may also be affected during the rainy period. It is observed that immediately after the cessation of rains, trees in wet evergreen forests tend to show phenomenal growth. It is possible that fresh roots are produced along with dense root hairs after the cessation of rains and root respiration is enhanced which is responsible for the enhanced, growth of trees. A systematic study on root hair adaptations of plants growing in wet evergreen forests would reveal interesting results.

Considering the important role of root hairs in root respiration of agricultural, horticultural crops and forest trees, the maintenance of proper soil aeration becomes a most important agronomic practice. This includes proper soil and water management including irrigation and soil working schedules. Improper soil aeration, soil compaction due to use of machines and lack of soil working in forest plantations would create oxygen stress in the soil and affect the root (hair) respiration. Continuous flood irrigation, as practised by the farmers, would choke the root hairs from proper aeration and thereby excessive soil moisture would either prevent development of fresh root hairs or degenerate the existing root hairs. In order to overcome low oxygen stress, plants tend to develop new roots towards more aerobic soils for which extra energy is utilized and may also bring down the shoot growth. Plants which suffer from root oxygen stress would also lose resistance against pathogens and succumb to root or shoot infections.

The new discovery on the respiratory role as the chief function of root hairs seems to have far-reaching implications in agriculture, forestry and horticulture. Root hairs now being projected as the main entry points for oxygen, the roots in flood irrigated soils may suffer from suffocation or lack of oxygen and, as a result, many of the growth metabolic activities of the crop plant may come to halt and would require some time to recover from the damage caused to the plant due to anaerobic environment created in the soil or may even kill the plant. In nodulating leguminous crops, excessive irrigation would kill root hairs and thereby reduce root nodulation, as root hairs are the only entry points for N_2-fixing microorganisms. Hence, restricted irrigation methods have proved to be more favourable for healthy crop growth and higher yields as they allow normal development and function of root hairs. Chemical fertilizers or ammonium nitrates might alter root hair development or, alternatively, block the synthesis of component/s involved in the perception or transduction of 'Nod' factors, thereby root hairs can no longer deform (Heidstra *et al.*, 1994). However, the information on the effect of chemical fertilizers on root hairs is prominently lacking.

In these days of decline in crop yields due to intensive cultivation practices, crop yields can be significantly enhanced without any extra inputs by promoting adequate root hair development and respiration through restricted irrigation methods. With the water thus saved, extra land can be brought under irrigation which will further increase food production. This is a better method of natural farming that farmers should adopt for increasing and sustaining food and wood production.

CHAPTER 8

Recommendations for Future Work

Looking into various implications of the respiratory role as the chief function of root hairs which are very sensitive to excessive soil moisture, following studies need to be carried out further to develop a suitable soil and water management practices for different crops.

1. Root hairs in aquatic plants need to be further examined with regard to their exact nature, cellular structure and mode of oxygen diffusion, as compared to the root hairs produced by land plants.

2. It would be worth investigating the adaptive significance of dark colored and persistent root hairs in *Anacardium occidentale* and light coloured persistent root hairs of *Ailanthus malabarica*.

3. The presence of cuticle on root hairs has been a controversial aspect. Whether there is really a cuticle on root hairs needs to be investigated, both among aquatic plants and terrestrial plants.

4. Studies can be conducted on the root hair responses of different agricultural, horticultural and forest crop plants against oxygen stress in different types of soils.

5. The tolerance range for root hairs in different crop plants against oxygen stress can offer scope for research.

6. One can develop optimum irrigation schedules and such irrigation methods which promote normal root hair development, respiration and growth.

7. Soil working is said to break the air continuity in soil, while zero tillage is said to maintain air continuity in soil pores, thereby enabling proper soil aeration. Therefore, there is a need to develop optimum soil tillage schedules to promote maximum root hair development and gaseous oxygen availability to roots.

8. Research can be done to standardize the threshold levels of soil aeration and soil moisture for different crops, which will promote optimum development of root hairs and provide better crop growth environment.

9. Investigations on the relationship between disease and pest resistance or susceptibility of crop plants with or without root hairs also hold promise.

10. It may also be necessary to find out the effect of chemical fertilizers and organic manure on root hair development, nodulation and respiration. The information on the effect of chemical fertilizers on root hairs is prominently lacking.

11. One can induce a greater number of root hairs on crop roots which do not have root hairs by biotechnological means or cultural methods to increase growth and yield of crops.

12. As there is great variation with regard to number, length and other parameters amongst cultivars, suitable cultivars should be selected to different sites. For example, cultivars having more number and lengthier root hairs are more suited to P deficient soils. Barley cv. Salka, with longer root hairs absorbed twice more P from rhizosphere soil than cv. Zita, with shorter root hairs, suggesting that P uptake from low-P

soils can be enhanced by selection of cereal cultivars having longer root hairs (Gahoonia and Nielsen, 1998).

13. Finally, suitable soil amendment material (such as coir pith, leaf litter, vermicompost, sawdust, sand, etc.) should be selected to improve soil aeration and root hair development. Probably, it may be required to carry out soil working with or without intercrops in forest plantations, which will help in root respiration, greater root growth and, consequently, increased absorption and translocation of water and minerals to the growing shoot.

Different energy dependent metabolic processes have been variously interpreted without a proper understanding of the basic adaptive mechanism of roots and root hairs for intake of free oxygen. For example, the so-called 'phosphate deficiency', as experienced by plant roots, may be due to soil anaerobiosis and damage to root hair–oxygen intake system rather than the actual deficiency of phosphorus in soil. Probably, if a better aerobic condition is created in soil and normal root hair production is promoted, automatically sufficient free oxygen supply is restored and phosphorus uptake is normalized either through root epidermis or root hairs. Several adaptations of plants grown under phosphate deficiency exhibit morphological adaptations by developing more root hairs. This may not be meant to enhance phosphorus uptake directly through root hairs but to enhance free oxygen diffusion into the root through root hairs to accelerate active uptake of phosphorus.

As root hairs assume prominence with regard to their chief role as respiratory 'gills', (nutrient uptake and other special roles in the infection of nitrogen-fixing microbes, VAM and also ectomycorrhizae being dependent functions), the future crop production research needs to focus on standardizing such agronomic practices as to promote better root hair development, so that it is possible to take full advantage of root hairs in growth respiration subsequently growth, vigour, disease resistance besides augmenting mineral uptake and to some extent, water absorption at no extra cost to the farmer.

CHAPTER 9

Summary

Roots and root hairs of higher plants are less studied as compared to above ground parts and, consequently, their physiological functions are less well understood. However, based on the work done so far on root hairs, the findings can be summarized as below:

(1) Root hairs are slender, usually thin-walled unicellular tubular evaginations from the outer wall of rhizodermal cells occurring in the root hair zone between the growing root tip and zone of active root elongation on a primary root.

(2) Root hairs contribute to anywhere between 75 and 80% of the surface area of the entire root system which possesses root hairs.

(3) There is a great variation in occurrence and form of root hairs. They are generally delicate structures and sensitive to dry or desiccated air or direct contact with liquid medium. Root hairs are ephemeral but with exceptions of *Anacardium occidentale*, *Ailanthus malabarica*, *Gleditsia triacanthos* which exhibit persistent root hairs. They are often unusually dark in colour and remain rigid even if exposed to dry air.

(4) Dimorphic root hairs are often reported in parasitic Scrophulariaceae, while none among Magnoliales has root hairs.

(5) Developmental pattern and number and size of root hairs are genetically controlled.

(6) Root hairs in some plants have been reported to exhibit a high degree of endo-polyploidy but in others, root hairs are found to maintain normal ploidy.

(7) Plants have been successfully regenerated from protoplasts released from root hairs.

(8) Root hairs have also been induced through microbial inoculations (using *Agrobacterium rhizogenes and penicillium bilaii*).

(9) The present author recognizes two types of root hairs in angiosperms: 'aquatic root hairs' (which occur among hydrophytes and are rare in nature e.g. *Alternathera sessilis*) and `aerial root hairs' (commonly occurring below ground in most terrestrial plants), which are different in cell wall structure and perhaps in other structural adaptations. In terrestrial plants, the outer layer of root hair cell has random micro-fibrils and forms a loose network and inner layer with axially-oriented micro-fibrils. Whereas, in aquatic plants, if at all there are any root hairs, the inner wall layer has helicoid micro-fibrils and outer wall layer has randomly oriented micro-fibrils. Among terrestrial plants, potato (*Solanum tuberosum*) is found to produce long and slender root hairs into the culture medium. Their structural and functional adaptations need to be studied.

Aquatic root hairs develop in surface water layers but are sensitive even to humid air, while the aerial root hairs are 'aero tropic' and are sensitive to direct wetting in liquid solutions. The aerial root hairs need optimal ambient humidity for their normal development and to maintain the turgidity. But both these types of root hairs need O_2-rich environment for their development and function as respiratory 'gills'. Aerial root hairs require a gaseous form of oxygen, while aquatic root hairs are capable of taking in dissolved oxygen.

(10) Almost all the previous workers have attributed water and mineral absorption as the main function of root hairs. However, phosphate transporters have been found enriched in the whole root epidermis, including root hairs of Pi deficient roots, confirming that root hairs are not necessarily involved in mineral uptake. Plant roots have also been shown to absorb water from other suberized parts of roots even in the absence of root hairs. In a simple experiment conducted by the present author, a drop of water was placed in the root hair zone in soybean and it was observed that root hairs in the submerged portion had completely disintegrated within a few hours. Interestingly, the drop of water was also not absorbed in the root hair zone. Thus, there are no convincing grounds for ascribing a specialized function of water and mineral uptake by root hairs. The exposed shoot surfaces can maintain normal respiration by utilizing the oxygen generated by photosynthesis and also oxygen diffused through stomates and lenticels, but the underground parts suffer from a more anaerobic condition compared to the shoot. In the absence of well-developed internal air passage system from shoot to root in land plants, the present author postulates that root hairs act as 'gills' in the subterranean root system, the chief role of root hairs in land plants being to serve as the entry points for diffusion of O_2 and to carry out respiration and all other metabolic processes are depdendent on the energy produced by root hairs. This new theory is well substantiated through various experimental and other evidences as summarized below:

(a) The protoplasm in the root hair tip is densely packed with cytoplasmic organelles, particularly with more mitochondria having maximum cristae development. Cytochrome oxidase present in mitochondria has a high affinity for oxygen. Mitochondria are represented in less numbers in non-root hair zones and evidently take less part in accepting oxygen from air. Perhaps the non-root hair zones, including deeper tissues in the living root, receive only dissolved oxygen or ATPs transported from the root hairs. The more rapid ATP or dissolved oxygen transport system can be expected from root hairs to other parts in roots with no lungs or haemoglobin to help tranport the oxygen. Such transport is not unreasonable, given the amount of cytoplasmic streaming that can be observed in root hairs. However, this transportation system of oxygen or ATPs from root hairs to other parts in roots needs further investigation.

(b) The greatest density of plasmodesmata has been recorded in the wall between the root hair and the first cortical cell, and in the radial walls between root hair and other hairless cells. Such a concentration of plasmodesmata is indicative of enhanced exchange of dissolved oxygen. Even free gaseous O_2 and CO_2 possibly move by simple diffusion across phospholipid bilayers.

(c) The cuticle is absent on root hairs. However, this requires further investigation among aquatic and terrestrial plants. Contrary to presumption that aerobic environment with oxygen would oxidize fatty acids to produce cuticle which would prevent root hair formation, the present author has proved that in many cases, root hair development was indeed more profuse and normal when roots were grown in aerobic or oxygen-rich environment but with sufficient relative humidity. Absence of cuticle in root hairs is a great advantage for the diffusion of oxygen; otherwise it would have formed a barrier as in other parts of roots which have a thin cuticle layer.

(d) There are 'pores' (unthickened areas in the cellulose wall wth thin microfibrillar network) throughout the entire wall of both young and old root hairs. They were found to increase in number and size near the base of the hair, suggesting that free oxygen diffusion to be greater towards the base of the root hair which is close to vacuole and has very little cytoplasm, unlike the root hair tip which has dense cytoplasm. The oxygen diffusion coefficient of cytoplasm is considered to be close to that of water. The diffusion of oxygen through water is said to be 1/10,000 of its diffusion rate in air [according to Drew (1979), oxygen diffusion in pure water is about three million times slower than in air]. There is an indication that only the tip of root hairs may aid in absorption of water and minerals. When 127.5 µm of the tip of root hair is immersed in a capillary, it will absorb, whereas the same root hair immersed at a depth of 255 µm will excrete. Thus, crops which are flood irrigated may indeed starve of water due to `physiological dryness'. Root hairs avoid direct contact or submersion in water or any liquid medium. This is shown through a simple experiment by placing a drop of water on dense root hairs in a soybean seedling root grown inside a glass petridish and constantly observed under a binocular research microscope. It was observed that the portion of root hair zone which was submerged in a drop of water resulted in a complete disintegration of root hairs within a few hours (4 h). Interestingly, the drop of water was also not absorbed by the root hair zone, disproving their involvement in water absorption.

(e) The fine bunches of roots with root hairs are usually concentrated in the upper soil horizons having closer proximity to oxygen source. Root hairs are often most conspicuously developed on roots growing in soil voids and in porous soils. When crops (like tomato) are grown under irrigation and the soil gets saturated with water, plants tend to produce new adventitious roots covered with dense root hairs along the exposed soil surface in order to survive out of soil anaerobiosis and to get free oxygen from air.

(f) The absence of root hairs as special 'respiratory gills' amongst submerged hydrophytes can be attributed to their development of efficient internal air passages to store oxygen and maintain normal respiration. Since root hairs are normally functional in the presence of gaseous oxygen, they do not simply develop under water as their presence serves no purpose. This is contrary to the earlier notion that since the entire plant body is immersed in water, hydrophytes do not need root hairs for absorption of water and minerals.

(g) However, there are a few hydrophytes which do have 'aquatic' root hairs and whose cell wall structure is different from the root hair cell wall of land plants. *Alternathera sessilis*, which is a semiaquatic or an emergent aquatic plant, produces dense and long root hairs only on the new long roots produced from the nodes just below the water level, while the nodal roots produced at deeper depths are without any root hairs. This is again a clear evidence to support that if at all there are any root hairs produced in completely-submerged hydrophytes, they are produced only closer to O_2 source, i.e. air or at upper reaches of water with more dissolved O_2. If their main function was not respiratory or 'absorption' of O_2, these root hairs would not have been produced nearer to air. The presence of helicoidal or axially-oriented micro-fibrils in the inner cell wall layers of aquatic root hairs seems to have some functional and adaptive significance in view of their development in liquid medium. Either aerial root hairs or aquatic root hairs, their chief function remains the same, i.e. intake of oxygen.

(h) Stomata in the root hair zone in *Pisum, Helianthus* and *Ornithopus* is an interesting phenomenon as they were found to be almost always open, usually without chloroplasts and were always without thickened cells at the border of the aperture. The present author feels that the stomata as observed in the root hair zone, may enable even greater diffusion of oxygen in addition to root hairs in highly-active tissues where root hairs alone may not suffice. Once the primary root branches out and increases the respiratory surface area, the need for stomata may be alleviated and hence such root branches are reported to degenerate. Observations of Tarkowska and Wacowska (1988) on the occurrence of stomata in root hair zone supports the present hypothesis that the root hair zone and especially the root hairs themselves are specialized for serving as 'gills'.

(i) Root hairs do not normally grow on roots that develop in tissue culture medium due to poor aeration and only those roots that grow outside the medium possess dense root hairs. Root hairs under submerged conditions are more or less functionless; hence they are absent on roots under submerged conditions. In *Arachis hypogea*, it has been observed that the root hairs were produced only above the nutrient solution level.

(j) A close relationship seems to exist between root hair, VAM and ectomycorrhizal associations. Root hairs are most essentially required for the entry of N_2-fixing microorganisms into the root system and form nodules, since root hairs provide adequate quantities of O_2 required for the microorganisms. Nitrogen fixation by Rhizobia in the infected root hair cells is directly dependent on a high rate of ATP synthesis through bacterial oxidative phosphorylation. The roles for leghaemoglobin present in the nodule have included O_2 binding, among others. Without leghaemoglobin, diffusion of O_2 through the dense nodule tissue would be completely inadequate to meet the ATP requirement. Therefore, in nitrogen-fixing legume nodules, oxygen is required in large amounts from aerobic respiration. It has been shown that nodules typically respire at about 4 times the rate of an equal biomass of roots. In case of soybean, the roots inside a sand medium or a soil having more porocity develop profuse root hairs and incidentally, abundant root nodules, whereas the roots which had grown into the nutrient solution or water rich soil environment were completely devoid of root hairs and nodules. This is further indicative of root hairs as the centres of O_2 diffusion and hence the microbial associations with them.

(11) There are several instances of increased yields in crops due to drip or sprinkler or alternate furrow irrigation methods as compared to flood or surface irrigation methods. The increased crop yields due to such restricted irrigation methods have been most commonly interpreted as due to continuous and optimum supply of water to roots without causing water stress during the growing period of crop plants. But the present author attributes it to increased or normal development of root hairs and consequent O_2 enrichment to roots enabling enhanced metabolic functions including energy dependent nutrient uptake. Of late, aeroponics is gaining more importance which involves growing plants by delivering a nutrient mist to the roots, because roots develop large amounts of root hairs and accordingly, roots are well oxygenated (in contrast with hydroponics) and plants grow healthier and yields are better (similar to drip irrigated crops). By following wrong agronomic and irrigation practices, nearly 75 to 90% of surface area (root hairs) which conduits oxygen to roots is blocked and to that extent, both maintenance and growth respiration are affected. In addition, the plants are forced to spend extra energy to produce new primary roots, thereby much of the energy which should be used for shoot growth would be diverted for new root formation. Hence, aeroponics, drip irrigation and such other restricted irrigation methods would give best results. This

would not only help in conserving water but also promote better soil aeration, together with better above ground biomass production.

In view of the new look at the chief role of root hairs as the 'respiratory gills', future research in crop production should be oriented towards providing a favourable soil environment to promote normal root hair development, oxygen supply through root hairs and increased crop growth and yields. This is a natural way of farming, aiming at increasing crop yields with reduced input costs in fertilization and irrigation, thereby with the water thus saved extra lands can be brought under irrigation and further increase food and wood production.

References

Abo El-Nil, M.M. and Hildebrandt, A.C. (1971). *Geranium* plant differentiation from anther callus. American Journal of Botany, **58**: 475.

Addoms, R.M. (1946). Entrance of water into suberized roots of trees. Plant Physiology, **21** : 109-111.

Adejare, G.O. and Coutts, R.H.A. (1981). Plant Cell Tissue and Organ Culture, **1**: 25-32. (cf. George and Sherrington, 1984).

Ahmed, M. and Cailloux, M. (1971). The effects of malonate on absorption of water by root hairs of *Avena sativa*. Canadian Journal of Botany, **49**: 521-528.

Ahmed, M. and Cailloux, M. (1972). Effects of some respiratory inhibitors on water flux in root hairs of *Avena sativa*. Canadian Journal of Botany, **50**: 575-579.

Aldrich, H.C., Ferl, R.J., Hills, M.H. and Akin, D.E. (1985). Ultrastructural correlates of anaerobic stress in corn roots. Tissue and Cell, **17**: 341-348.

Alessa, L., Earnhart, C.G., Cole, D.N. (eds.) (2000). Effects of soil compaction on root and root hair morphology: implications for composite rehabilitation. Wilderness Science in a time of change conference. Vol. 5: Wilderness ecosystems, threats, and management, Missoula, Montana, USA, 23-27 May 1999. Proceedings, Rocky Mountain Research Station, USDA Forest Service, 2000, **5:** 99-104.

Alisa and Soran (1985). The effect of higher and lower temperature on protoplasmic streaming of wheat (*Triticum aestivum*). Rev. Roum Biol. Ser. Biol. Veg., **30:** 131-136.

Alpi, A. and Beevers H. (1983). Effects of O_2 concentration on rice seedlings. Plant Physiology, **71:** 30-34.

Amstel, A.N.M. van, Derksen, J. and Van-Amstel, A.N.M. (1993). The complex helical texture of the secondary cell wall of *Urtica dioica* root hairs is not controlled by microtubules: a quantitative analysis. Acta-Botanica-Neerlandica, **42**(2): 141-151.

Arber, A. (1934). The Gramineae. Cambridge University Press, Cambridge.

Ardourel, M., Demont N., Debelle F., Maillet F., De Billy F., Prome J., Denarie J. and Truchet G. (1994). *Rhizobium meliloti* lipooligosaccharide nodulation factors: Different structural requirements for bacterial entry into target root hair cells and induction of plant symbiotic developmental responses. The Plant Cell, **6:** 1357-1374.

Armstrong, W. (1979). Aeration in higher plants: In: Advances in Botanical Research, 7: 226-332. ed. H.W. Woolhouse, Academic Press, London.

Armstrong, W. (1982). Waterlogged soils. In: J.R. Etherington, ed. Environment and Plant Ecology. John Wiley & Sons, Chichester, UK, pp. 290-330.

Armstrong, W. and Beckett, P.M. (1985). Root aeration in unsaturated soil: A multi-shelled mathematical model of oxygen diffusion and distribution with and without sectoral water-soil blocking of the diffusion path. New Phytologist, **100:** 293-311.

Armstrong, W. and others (1990). Modeling root aeration. In: Plant life under oxygen deprivation; Ed. M.B. Jackson, D.D. Davies and H. Lambers. SPB Academic Publishing, The Netherlands.

Armstrong, W., Beckett, P.M., Justin S.H.F.W., Lythe, S. (1991). Modeling and other aspects of root aeration by diffusion. In: Jackson, M.B., Davies, D.D., Lambers, H. (eds) 'Plant life under oxygen deprivation', The Hague: SPB Academic, 267-282.

Arteca, R.N. and Dong, C.N. (1981). Increased photosynthetic rates following gibberellic acid treatments to the roots of tomato plants. Photosynth. Res. **2**: 243-249.

Arteca, R.N., Poovaiah, B.W. and Smith, O.E. (1979). Changes in carbon fixation, tuberization and growth induced by CO_2 applications to the root zone of potato plants. Science, **205**: 1279-1280.

Aslyng, H.C. (1986). Water balance and crop production. In: Agricultural Water Management (Eds.) AL M. Van Wijk and J. Wesseling. pp.129-139. A.A. Balkema, Rotterdam.

Atwell, B.J. and Greenway, H. (1987). Carbohydrate metabolism of rice seedlings grown in oxygen deficient solution. J. Expt. Bot., **38**: 466-478.

Atwell, B.J., Thomson, C.J., Greenway, H., Ward, G. and Waters, I. (1985). A study of the impaired growth of roots of *Zea mays* seedlings at low oxygen concentrations. Plant Cell Environment, **8**: 179-188.

Bagyaraj, D.J., Manjunath, A. and Patil, R.B. (1979). Occurrence of vesicular-arbuscular mycorrhizas in some tropical aquatic plants. Transactions of British Mycological Society, **72**: 164-167.

Baird, W.V. and Riopel, J.L, (1983). Experimental studies on the attachment of haustoria in *Agalinis purpurea* to a host. Protoplasma, **118**: 206-218.

Baird, W.V. and Riopel, J.L. (1985). Surface characteristics of root and haustorial hairs of parasitic Scrophulariaceae. Botanical Gazette, **146**: 63-69.

Bakhuizen, R. (1988). The plant cytoskeleton in the Rhizobium-legume symbiosis. Ph.D. thesis. Leiden University, The Netherlands (cf. Heidstra *et al.*, 1994).

Barber, C.A. (1906a). Selections from Reports and Notes on spike disease in sandal. Mysore Forest Dept.

Barber, C.A. (1906b). Memoirs of Department of Agriculture, India, Botanical Series Bulletin, **3** (1917).

Barber, D.A., Ebert, M. and Evans, N.T.S. (1962). J. Exptl. Bot., **13**: 397-403.

Barnola, L.G. and Montilla, M.G. (1997). Vertical distribution of mycorrhizal colonization, root hairs, and belowground biomass in three contrasting sites from the tropical high mountains Merida, Venezuela. Arctic and Alpine Research, **29**(2): 206-212.

Bauer, W.D. (1981). Infection of legumes by Rhizobia. Ann. Rev. Plant Physiol., **32**: 407-449.

Baylis, G.T.S. (1970). Root hairs and phycomycelous mycorrhizas in phosphorus deficient soil. Plant & Soil, **33**: 713-771.

Baylis, G.T.S. (1972). Minimum levels of available phosphorus for non-mycorrhizal plants. Plant & Soil, **36**: 233-234.

Baylis, G.T.S. (1975). The magnolioid mycorrhiza and mycotrophy in root systems derived from it. In: Endomycorrhizas (Ed. by F.E. Sanders, B. Mosse and P.B. Tinker), pp. 373-389. Academic Press, New York.

Belford, D.S. and Preston, R.D. (1961). The structure and growth of root hairs. Journal of Experimental Botany, **12**: 157-168.

Bergersen, F.J. (1962). The effects of partial pressure of O_2 upon respiration and N_2 fixation by soybean root nodules. Journal of General Microbiololgy, **29**: 113.

Bergersen, F.J. (1971). Biochemistry of symbiotic nitrogen fixation in legumes. Annual Review of Plant Physiology, **22**: 121.

Berlyn, G.P. (1967). The structure of germination in *Pinus lambertiana* Dougl. Bulletin No.71. Yale University School of Forestry, New Haven, Conn.

Berlyn, G.P. (1972). Germination and morphogenesis. In: T. Kozlowski (ed.), Seed biology. Academic Press, New York, pp. 223-311.

Berry, A.M., McIntyre, L. and McCully, M.E. (1986). Fine structure of root hair infection leading to nodulation in the *Frankia-Alnus* symbiosis. Canadian Journal of Botany, **64**: 292-305.

Berry, A.M. and Torrey, J.C. (1983). Root hair deformation in the infection process of *Alnus rubra*. Can. J. Bot., **61**: 2863-2876.

Berry, A.M., Torrey, J.G. and McCully, M.C. (1983). The fine structure of the root hair wall and surface mucilage in the actinorrhizal host, *Alnus rubra*. Cellular Molecular Biology & Plant Stress, **12**: 319-327.

Bertl, A. and Felle, H. (1985). Cytoplasmic pH of root hair cells of Sinopsis alba recorded by a pH-sensitive micro-electrode: Does Fusicoccin stimulate the proton pump by cytoplasmic acidification? Journal of Experimental Botany, **36**: 1142-1149.

Bhaskar, V., Berlyn, G.P. and Conolly, J.H. (1993). Root hairs as specialized respiratory cells—A new hypothesis. Journal of Sustainable Forestry. **1** (2): 107-125.

Bhat, K.K.S. and Nye, P.H. (1973). Diffusion of phosphate to plant roots in soil. I. Quantitative auto radiography of the depletion zone. Plant & Soil, **38**: 161-175.

Bhat, K.K.S. and Nye, P.H. (1974). Diffusion of phosphate to plant roots in soil. II. Uptake along the root at different times and the effect of different levels of phosphorus. Plant & Soil, **41**: 365-382.

Bhat, K.K., Nye, K.K. and Baldwin, J.P. (1976). Diffusion of phosphate to plant roots in soil. IV. The concentration distance profile in the rhizosphere of roots with root hairs in low-P soil. Plant & Soil, **44**: 63-72.

Bidel, Lec P.R., Renault, P., Pages, L. and Riviere, L.M. (2001). An improved method to measure spatial variation in root respiration: application to the taproot of a young peach tree *Prunus persica*. Agronomie **21**: 179-192.

Bogar, G.D. and Smith, F.H. (1965). Anatomy of seedling roots of *Pseudotsuga menziesii*. American Journal of Botany, **52**: 720-729.

Bole, J.B. (1973). Influence of root hair in supplying soil phosphorus to wheat. Canadian Journal of Soil Science, **53**: 169-175.

Bonnett, H.T. and Newcomb, E.H. (1966). Coated vesicles and other cytoplasmic components of growing root hairs of radish. Protoplasma, **62**: 59-75.

Boon, F.R. and Veen, B.W. (1993). Mechanisms of crop responses to soil compaction. In: Soil Compaction in Crop Production. Eds. B.D. Soane and C. van Vuwerkerk. Elsevier, Amsterdam.

Bordonaro, J.L. and Curtis, W.R. (2000). Inhibitory role of root hairs on transport within root culture bioreactors. Biotechnology & Bioengineering, **70**(2): 176-186.

Bothe, H., Korsgen, H., Lehmacher, T. and Hundeshagen, B. (1992). Differential effects of *Azospirillum*, auxin and combined nitrogen on the growth of the roots of wheat. Symbiosis-Rehovot., **13**(1-3): 167-179.

Bowen, G.D. and Rovira, A.D. (1969). The influence of microorganisms on growth and metabolism of plant roots. In: Root Growth (Ed. Whittington, W.J.). Butterworths, London.

Bowen, G.D. and Rovira, A.D. (1976). Microbial colonization of plant roots. Annual Review of Phytopathology, **14**: 121-144.

Bradford, K.J. and Young, S.F. (1981). Physiological responses of plants to water logging. Horticultural Science, **16**: 25-30.

Brauer, D., Uknalis, J., Triana, R., Shachar, H.Y., Tu, S. and Tu, S.I. (1997). Effects of bafilomycin A1 and metabolic inhibitors on the maintenance of vacuolar acidity in maize root hair cells. Pl. Physiol. **113**(3): 809-816.

Brewin, N.J. and Kardailsky, I.V. (1997). Legume lectins and nodulation by Rhizobium. Trends in Plant Science, 2(3): 92-98.

Broek, A. Vande, Michiels, J., Gool, A. van, Vanderleyden, J., Vande-Broek, A., Van-Gool, A. (1993). Spatial-temporal colonization patterns of *Azospirillum brasilense* on the wheat root surface and expression of the bacterial nifh gene during association. Molecular-Plant-Microbe-Interactions, 6(5): 592-600.

Brook, P.J. (1952). Mycorrhiza of *Perneltya macrostigma*. New Phytol., 51: 388-397.

Brown, W.H. (1913). The relation of the substratum to the growth of *Elodea*. Philippine Journal of Science C. Botany, 8: 1-20.

Buyanovsky, G.A. and Wagner, G.H. (1983). Annual cycles of CO_2 level in soil air. Soil Science Society American Journal, 47: 1139-1145.

Cailloux, M. (1943). Mesure quantitative de l"ean absorbee par un seul poil radiculaire. Annales de L'Association Canadienne Francais Pour L"Advancement des Sciences, 10, 84.

Cailloux, M. (1950). Observations sur l'exsudation par des poils radiculaires. Ann. ACFAS 16: 158-162.

Cailloux, M. (1953). Sur la localisation de lae"gion par on l'eau pentre dans les poils radiculaires. Review of Canadian Biology, 11, 505-508.

Cailloux, M. (1972). Metabolism and the absorption of water by root hairs. Canadian Journal of Botany, 50: 557-573.

Cailloux, M. (1974). Metabolism and the absorption of water by root hairs. In: Structure and function of primary root tissues (J. Kolek, ed.), Veda, Bratislava, Czechoslovakia, pp. 315-322.

Callaham, D. and Torrey, J.G. (1977). The structural basis for infection of root hairs of *Trifolium repens* by *Rhizobium*. Canadian Journal of Botany, 59: 1647-1664.

Callaham, D., Newcomb, W., Torrey, J.G. and Peterson, R.L. (1979).

Root hair infection in actinomycete-induced and nodule initiation in *Casuarina, Myrica* and *Comptonia* (Myricaceae). Canadian Journal of Botany, **55**: 1647-1664.

Cambell, R. and Drew, M.C. (1983). Electron microscopy of gas space (aerenchyma) formation in adventitious roots of *Zea mays* L. subjected to oxygen shortage. Planta, **157**: 350-357.

Cao-XiaoFeng, Linstead, P., Berger, F., Kieber, J., Dolan, L. and Cao-X.F. (1999). Differential ethylene sensitivity of epidermal cells is involved in the establishment of cell pattern in the *Arabidopsis* root. Physiologia Plantarum, **106**(3): 311-317.

Caradus, J.R. (1979). Selection for root hair length in white clover (*Trifolium repens* L.). Euphytica, **28**: 489-494.

Carpenter, J.R. and Mitchell, C.A. (1980). Root respiration characteristics of flood tolerant and intolerant tree species. Journal of American Society of Horticultural Science, **105**: 684-687.

Ceremonia, H., Debelle, F., Fernandez, M.P. and Dawson, J.O. (eds.). (1999). Structural and functional comparison of Frankia root hair deforming factor and rhizobia Nod factor. Canadian Journal of Botany, **77**(9): 1293-1301.

Chaboud, A. (1983). Isolation, purification and chemical composition of maize root cap slime. Plant & Soil, **73**: 395-402.

Chandler, M.R. (1978). Some observations on the infection of *Arachis hypogea* L. by *Rhizobium*. Journal of Experimental Botany, **29**: 749-755.

Charlton, W.A. (1978). Some aspects of the roots of aquatic monocotyledons. Hellobiae Newsl., **2**: 9-11.

Chaubal, R., Sharma, G.D. and Mishra, R.R. (1982). Vesicular-arbuscular mycorrhiza in subtropical aquatic and marshy plant communities. Proceedings of Indian Academy of Science (Plant Science), **91**: 69-77.

Chauvel, A., Grimaldi, M., Tessier, D. (1991). Changes in soil pore-space distribution following deforestation and revegetation: an example from the Central Amazon basin. Forest Ecology & Management, **38**: 259-271.

Chilvers, G.A. and Proyor, L.D.(1965). The structure of eucalypt mycorrhizas. Austral. J. Bot., **13**: 245-259.

Chirkova, T.V. and Gutman, T.S.(1972). Physiological role of branch lenticels in willow and poplar under conditions of root anaerobiosis. Soviet Plant Physiology, **19**: 289-295.

Christensen, P.B., Revsbeck, N.P. and Sand-Jensen, K. (1994). Micro sensor analysis of oxygen in the rhizosphere of the aquatic macrophyte *Littorella uniflora* (L.) Ascherson. Plant Physiol., **105**: 847-852.

Christodoulakis, N.S. and Psaras, G.K. (1987). Stomata on the primary root of *Ceratonia siliqua*. Annals of Botany, **60**: 295-297.

Clarke, A.E., Anderson, R.L. and Stone, B.A. (1979). Form and function of arabinoglactans and arabinoglactan proteins. Phytochemistry, **18**: 521-540.

Clarkson, D.T. (1981). Nutrient interception and transport by root systems. In: Physiological processes limiting plant productivity (ed. C.B. Johnson), pp. 308-330. Butterworths.

Clarkson, D.T. and Hanson, J.B. (1980). The mineral nutrition of higher plants. Annual Review of Plant Physiology, **31**: 239-298.

Clarkson, D.T., Robards, A.W., Stephens, J.E. and Stark, M. (1987). Suberin lamellae in the hypodermis of maize (*Zea mays*) roots: development and factors affecting the permeability of hypodermal layers. Plant, Cell and Environment, **10**: 83-94.

Clayton, J.S. and Bagyaraj, D.J. (1984). Vesicular-arbuscular mycorrhizas in submerged aquatic plants of New Zealand. Aquatic Botany, **19**: 251-262.

Clemens, J., Kirk, A.M. and Mills, P.D. (1978). The resistance to waterlogging of three Eucalyptus species, effect of flooding and on ethylene releasing growth substances on *E. robusta, E. grandis* and *E. saligna*. Oecologia, **34**: 125-131.

Cocking, E.C. (1985). Protoplasts from root hairs of crop plants. Biotechnology, **3**: 1104-1106.

Connolly, J.H. and Berlyn, G.L. (1996). Cytochemical assay for differential respiratory activity in roots and root hairs. Biotehmic and Histochemistry, **71**: 197-201.

Conway, V.M. (1937). Studies in the autoecology of *Cladium mariscus* R. Br. III. The aeration of subterranean parts of the plant. New Phytologist, **36**: 64-96.

Cormack, R.G.H. (1937). The development of root hairs by *Elodea canadensis*. New Phytologist, **36**: 19-26.

Cormack, R.G.H. (1949). The development of root hairs in angiosperms. Botanical Review, **15**: 583-612.

Cormack, R.G.H. (1962). Development of root hairs in angiosperms. II. Botanical Review, **28**: 446-464.

Coult, D.A. and Vallance, K.B. (1951). Observations on the gaseous exchanges which take place between *Menyanthus trifoliata* L. and its environment. I. Journal of Experimental Botany, **2**: 212-222.

Coult, D.A. and Vallance, K.B. (1958). Observations on the gaseous exchanges which take place between *Menyanthus trifoliata* L. and its environment. II. J. Exp. Bot., **9**: 384-402.

Coulter, J.M., Barnes, C.R. and Cowles, H.O. (1911). Textbook of Botany. Vol. 2. Ecology. American Book Co.

Coupin, H. (1919). Sur l'absorption de sels mineraux par le summet de la racing. C.R. Hebd. Seances Acad. Sci. Ser.D. Sci. Nat. (Paris), **169**: 242-245.

Coutts, M.P. and Philipson, J.J. (1978). Tolerance of tree roots to water logging II. Adaptation of sitka spruce and lodgepole pine to water logged soil. New Phytologist, **80**: 71-77.

Crawford, R.M.M. (1971). Some metabolic aspects of ecology. Transactions of the Botanical Society, Edinburgh, **41**: 309-322.

Crawford, R.M.M. (1975). Metabolic adaptations to anoxia in plants and animals. Proceedings of the 12th International Botanical Congress, Leningrad.

Crawford, R.M.M. (1976). Tolerance of anoxia and the regulation of glycolysis in tree roots. In: Tree physiology and yield improvement (ed. by M.G.R. Cannel and F.T. Last), Academic Press, London.

Curran, N., Cole, M. and Allaway, W.G. (1986). Root aeration and respiration in young mangrove plants (*Avicennia marina*). Journal of Experimental Botany, **37**: 1225-1233.

Cutter, E.G. (1970). Plant anatomy: Experiment and interpretation, Part 1, Cells and tissues. Edward Arnold, London.

Cutter, E.G. and Feldman, L.J. (1970a). Trichoblasts in *Hydrocharis*: I. Origin, differentiation, dimensions and growth. American Journal of Botany, **57**: 190-201.

Cutter, E.G. and Feldman, L.J. (1970b). Trichoblasts in *Hydrocharis*: II. Nucleic acids, proteins and a consideration of cell growth in relation to endopolyploidy. American Journal of Botany, **57**: 202-211.

Dacey, J.W.H. (1980). Internal winds in water lillies: An adoption for life in anaerobic sediments. Science, **210**: 1017.

Dale, H.M. (1951). Carbon dioxide and root hair development in *Anacharis* (*Elodea*). Science, **114**: 438-439.

Darnell, J., Lodish, H. and Baltimore, D. (1990). Molecular cell biology. 2nd ed., Scientific American Books, N.Y. p.532.

Davies, W.J. (1986). Transpiration and water balance of plants. In: Plant physiology: a treatise Vol. IX. Water and solutes in plants (ed. by F.C. Steward), pp. 49-154. Academic Press.

Dawes, C.J. and Bowler, P. (1959). Light and electron microscope studies of the cell wall structure of the root hairs of *Raphanus sativus*. American Journal of Botany, **46**: 561-565.

Dayanandan, P., Barnabas, A.D., Jayakumar, P.S. and Christopher, J. (1986). Observations on peculiar air passages in the stems of *Gloriosa superba* and *Iphigenia indica*. Current Science (Bangalore), **55**: 235-238.

Dazzo, F.B., Hollingsworth, R., Abe, M., Smith, K.B., Welsch, M., Morris, P.J., Hollingsworth, S.P., Salzwedel, J.L., Castillo, R.M. (1987). Rhizobium trifolii polysaccharides, oligosaccharides and other metabolites affecting development and symbiotic infection of clover root hairs. In: G.L. Steffens and T.S. Rumsey (eds.), Biomechanisms regulating growth and development: Keys to progress. Vol. XII USDA Agri. Res. Center Symp., USDA, Beltsville, MD, pp. 343-355.

Dazzo, F.B., Truchet, G.L., Hollingsworth, R.I., Hrabak, E.M., Pankratz, H.S., Philip-Hollingsworth, S., Salzwedel, J.L., Chapman, K., Appenzeller, L., Sqartini, A., Gerhold, D. and Orgambide, G. (1991). Rhizobium lipopolysaccharide modulates infection thread development in white clover root hairs. Journal of Bacteriology, **173** (17): 5371-5384.

Debaene-Gill, S.B., Allen, P.S. and Gardner, J.S. (1994). Morphology of the perennial eyegrass (*Lolium perenne*; Poaceae) coleorhiza and emerging radicle with continuous or discontinuous hydration. American Journal of Botany, **81** (6): 739-744.

Debergh, P. and Maene, L. (1981). Scientia Horticulture, **14**: 335-345 (cf. George and Sherrington, 1984).

Denarie, J. and Cullimore, J. (1993). Lipo-oligosaccharide nodulation factors: a new class of signaling molecules mediating recognition and morphogenesis. Cell, **74**: 951-954.

Derksen, J., Jeucken, G., Traas, J.A., and van Lammeren, A.A.M. (1986). The micro tubular skeleton in differentiating root tips of *Raphanus sativus* L. Acta Botanica Neerlandica, **35**: 223-231.

Ditengou, F.A., Beguiristain, T. and Lapeyrie, F. (2000). Root hair elongation is inhibited by hypaphorine, the indole alkaloid from the ectomycorrhizal fungus *Pisolithus tinctorius*, and resored by indole-3-acetic acid. Planta, **211**(5): 722-728.

Devlin, R.M. and Witham, F.H. (1983). Plant Physiology. 4th ed. Willard Grant Press, Boston.

Diaz, C.L., van Spronsen, P.C., Bakhuuizen, R., Logman, G.J.J., Lugtenberg, E.J.J. and Kijne, J.W. (1986). Correlation between infection by *Rhizobium leguminosarum* and lectin on the surface of *Pisum sativum* L. roots. Planta, **168**: 350-359.

Dittmer, H.J. (1937). A quantitative study of the roots and root hairs of a winter rye plant (*Secale cereale*). American Journal of Botany, **24**: 417-420.

Dittmer, H.J. (1949). Root hair variations in plant species. American Journal of Botany, **36**: 152-155.

Dong, H.S., Wang, J.S. and Fang, C.T. (1993). Reversibility, irreversibility, and other properties of adsorption of *Erwinia carotovora* subsp. *cartovora* to Chinese cabbage root hairs. Acta Phytopathologica Sinica, **23** (3): 275-280.

Drew, M.C. (1979). Plant response to anaerobic conditions in soil and solution culture. Current Advances in Plant Sciences, **36**: 124.

Drew, M.C. (1987). Function of root tissues in nutrient and water transport. In: Root development and function (ed. by P.J. Gregory *et al.*), Cambridge Univ. Press.

Drew, M.C. (1992). Soil aeration and plant root metabolism. Soil Science, **154**: 259-268.

Drew, M.C., Saglio, P.H. and Pradet, A. (1985). Larger adenylate energy charge and ATP/ADP ratios in aerenchymatous roots of *Zea mays* L. in anaerobic media as a consequence of improved internal oxygen transport. Planta, **165**: 51-58.

Ekdahl, I. (1953). Studies on the growth and the osmotic conditions of root hairs. Sym. Bot. Ups., **11**: 1-83.

Ellmore, G.S. (1981). Root dimorphism in *Ludwigia peploides* (Onagraceae: Structure and gas content of mature roots. American Journal of Botany, **68**: 557-568.

Emons, A.M.C. (1985). Plasma-membrane rosette in root hairs of *Equisetum hyemale*. Planta, **163**: 350-359.

Emons, A.M.C. and van Maaren, N. (1987). Helical cell wall texture in root hairs. Planta (Berlin), **170**: 145-151.

Erdmann, B., and Wiedenroth, E.M. (1986). Changes in the root system of wheat seedlings following root anaerobiosis. II. Morphology and anatomy of evolution forms. Annals of Botany, **58**: 607-616.

Erdmann, B., and Wiedenroth, E.M. (1988). Changes in the root system of wheat seedlings following root anaerobiosis: III. Oxygen concentration in the roots. Annals of Botany, **62**: 227-286.

Erdmann, B., Hoffman, P. and Wiedenroth, E.M. (1986). Changes in the root system of wheat seedlings following root anaerobiosis. I. Anatomy and respiration in *Triticum aestivum* L. Annals of Botany, **58**: 597-605.

Erickson, L.G., (1946). Growth of tomato roots as influenced by oxygen in the solution. American Journal of Botany, **33**: 551-561.

Evans, N.T.S. and Ebert, H. (1960). Radioactive oxygen in the study of gas transport down the root of *Vicia faba*. J. Exptl. Bot., **11**: 246-257.

Ewell, K.C., Wendell, Jr. P.C. and Gholz, H.L. (1987). Soil carbon dioxide evolution in Florida (USA) slash pine plantations: II. Importance of root respiration. Canadian Journal of Forestry Research, **17**: 330-333.

Ewens, M. and Leigh, R.A. (1985). The effect of nutrient solution composition on the length of root hairs of wheat (*Triticum aestivum* L.). Journal of Experimental Botany, **36**: 713-724.

Fahn, A. (1982). Plant Anatomy. Pergamon Press, Oxford.

Farr, C.H. (1928a). Studies on the growth of root hair in solutions V. Root hair elongation as an index of root development. American Journal of Botany, **15**: 103-113.

Farr, C.H. (1928b). Studies on the growth of root hairs in solutions VI. Structural responses to toxic pH and molar concentrations of calcium chloride. American Journal of Botany, **15**: 171-178.

Farr, C.H. (1928c). Root hairs and growth. Quarterly Review of Biology, **3**: 343-376.

Felle, H. (1987). Polar transport and pH control in *Sinapsis alba* root hairs: a study carried out with double-barreled pH microelectrodes. Journal of Experimental Botany, **38**: 340-354.

Filippenko, V.N. (1981). Determination of the direction of rhizodermal cells differentiation in corn. Soc. J. Dev. Biol., **11**: 189-196.

Fisher, R.F. and Long, S.R. (1992). Rhizobium-plant signal exchange. Nature, **357**: 655-660.

Fohse, D., Claassen, N., Jungk, A. (1991). Phosphorus efficiency in plants. Plant & Soil, **132**: 261-272.

Gahoonia, T.S. and Nielsen, N.E. (1998). Enhancing soil phosphorus use with longer root hairs of cereal cultivars. Kungl.-Skogs-och-Lantbruksakademiens-Tidskrift, **137**(7): 131-136.

Gahoonia, T.S., Care, D. and Nielsen, N.E. (1997). Root hairs and phosphorus acquisition of wheat and barley cultivars. Plant & Soil, **191**(2): 181-188.

Gassman, W. and Schroeder, J.I. (1993). Inward-rectifying K+ channel currents in root hairs of wheat (Abstr.). Plant Physiology (Suppl.), **102**: 150.

Gay, C., Corbineau, F. and Come, D. (1991). Effects of temperature and oxygen on seed germination and seedling growth in sunflower (*Helianthus annuus* L.). Environmental and Experimental Botany, **31** (2): 193-200.

Geisler, G. (1965). The morphogenetic effect of oxygen on roots. Plant Physiology, **40**: 85-88.

George, E.F. and Sherrington, P.D. (1984). Plant propagation by tissue culture. Exegetics Ltd. England.

Gerrit, S., Kijne, J.W., Lugtenberg, B.J.J. (1986). Correlation between extra-cellular fibrils and attachment of *Rhizobium leguminosarum* to pea root hair tips. Journal of Bacteriology, **168**: 821-827.

Ghelue, M. van, Lovaas, E., Ringo, E., Solheim, B., Van Ghelue, M. Berry, A.M. (eds.) (1995). Early interactions between *Alnus glutinosa* and Frankia strain Ar13. Production and specificity of root hair deformation factor(s). Proceedings, Tenth Intl. Conf. on Frankia and Actinorhizal plants, Davis, California, USA, 6-11 August. Physiologia Plantarum, **99**(4): 579-587.

Gill, C.J. (1970). The flooding tolerance of woody species—a review. Forestry Abstract, **31**: 671-688.

Gilroy, S. and Jones, D.L. (2000). Through form to function: root hair development and nutrient uptake. Plant and Soil, **211**(2): 269-281.

Graterol, Y.E., Eisenhauer, D.E. and Elmore, R.W. (1993). Alternate-furrow irrigation for soybean production. Agri. Water Management, **24**: 133-145.

Greenland, D.J. (1979). The physics and chemistry of the soil-root interface: some comments. In: The Soil-Root Interface, Eds. Harley, J.L. and Russel, R.S.. Academic Press, London.

Greenwood, D.J. (1967). Studies on the transport of oxygen through the stems and roots of vegetable seedlings. New Phytol., **66**: 337-347.

Guafa, W., Peoples, M.B., Herridge, D.F. and Rerkasem, B. (1993). Nitrogen fixation, growth and yield of soybean grown under saturated soil culture and conventional irrigation. Field Crops Research, **32**: 257-268.

Guhmann, H. (1924). Variation in the root system of the common everlasting *Gnaphalium, Polycephalum*. Ohio Journal of Science, **24**: 199-207.

Gulden, R.H. and Vessey, J.K. (2000). *Penicillium bilaii* inoculation increases root-hair production in field pea. Canad. J. Pl. Sci., **80**(4): 801-804.

Gullan, P.K. (1975). Vegetation at Cranbourne vol.2. Ph.D. Thesis, Monash Univ., Victoria (cf. Lamont, 1982).

Hameed, K.M. and Foy, C.L. (1991). Observations on the primary haustorium formation by germinated *Orobanche ramosa* seeds in relation to suscept and non-suscept plants. Proc. 5th Intl. Symp. of Parasitic Weeds. Nairobi (Kenya). pp. 36-42.

Harley, J.L. (1973). Symbiosis in the ecosystem. J. Nat. Sci. Council Sri Lanka, **1**: 31-48.

Harris, D.G. and van Bavel, C.H.M. (1957a). Growth and yield and water absorption of tobacco plants as affected by the composition of the root atmosphere. Agronomic Journal, **49**: 11-14.

Harris, D.G. and van Bavel, C.H.M. (1957b). Root respiration of tobacco, corn and cotton plants. Agronomic Journal, **49**: 182-184.

*Harris, R.E. and Stevenson, J.H. (1979). Comb. Proc. Int. Plant Prop. Soc., **29**: 95-108. (cf. George and Sherrington, 1984).

Harris, W.M. (1979). Ultra-structural observations on the trichoblasts of *Equisetum*. American Journal of Botany, **66**: 673-684.

Hawkins, J.C. (1962). The effects of cultivation on aeration, drainage, and other soil factors important in plant growth. J. Sci. Food Agric., **13** : 386-391.

Hayman, D.S. (1978). Endomycorrhizas. In: Interaction between non-pathogenic soil micro-organisms and plants. Eds. Dommergues, Y.R. and Krupa, S.V., Elsevier Sci. Publ. Co., Amsterdam, pp. 400-442.

Hayward, H.E., Blair, W.M. and Skaling, P.E. (1942). Device for measuring entry of water into roots. Botanical Gazette, **104**: 152-160.

Healy, M.T. and Armstrong, W. (1972). The effectiveness of internal oxygen transport in a mesophyte (*Pisum sativum* L.). Planta (Berlin), **103**: 302-309.

Heidstra, R., Geurts, R., Franssen, H., Spaink, H.P., Kammen, Ab van, and Bisseling T. (1994). Root hair deformation activity of nodulation factors and their fate on *Vicia sativa*. Plant Physiol., **105**: 787-797.

Hermina, N. and Reporter, M. (1977). Root hair cell enhancement in tissue cultures from soybean roots: a useful model system. Plant Physiology, **59**: 97-102.

Hearn, A.B. and Constable, G.A. (1981). Irrigation for crops in a sub-humid environment I. Stress day analysis for soybeans and an economic evaluation of strategies. Irrigation Science, **3**: 1-15.

Hesse, H. (1904). Beitrage zur morphologie und biologie der wurzelhaare. Thesis, University of Jena. (cf. Hofer, 1991).

Hochmuth, G.J. (1986). A gene affecting tomato root hair production in solution culture. Rep. Tomato Genet. Coop., **36**: 4-6.

Hofer, Rose-Marie (1991). Root hairs. In: Plant roots—the hidden half. Ed. Yoav Waisel *et al.*, Marcel Dekker Inc., N.Y. pp. 129-148.

Hook, D.D. (1984). Adaptations to flooding with fresh water. In: T.T. Kozlowski (ed.), Flooding and plant growth. Acad. Press, N.Y., pp. 265-294.

Hook, D.D., Brown, C.L.L. and Kormanik, P.P. (1971). Inductive flood tolerance in swamp tupelo [*Nyssa sylvatica* var. *biflora* (Walt.) Sarg.]. Journal of Experimental Botany, **22**: 78-89.

Horan, D.P. and G.A. Chilvers (1990). Chemotropism—the key to ectomycorrhizal formation? New Phytol., **116**: 297-301.

Hosner, J.F. and S.G. Boyce (1962). Relative tolerance to water saturated soil of various bottomland hardwoods. Forest Science, **8**: 180-186.

Hostrup, O. and Wiegleb, G. (1991). Anatomy of leaves of submerged and emergent forms of *Littorella uniflora* (L.) Ascherson. Aquat. Bot., **39**: 195-209.

Huss-Danell, K. (1997). Actinorhizal symbioses and their N2 fixation. New Phytol., **136**(3): 375-405. (Tansley Review No. 93)

Itoh, S. and Barber, S.A. (1983). A numerical solution of whole plant nutrient uptake for soil-root systems with root hairs. Plant & Soil, **70**: 403-413.

Jackson, M.B., Abbott, A.J., Belcher, A.R. and Hall, K.C. (1987). Gas exchange in plant tissue cultures. In: Jackson *et al.*, Advances in the chemical manipulation of plant tissue cultures. Monogr. 16, British pl. Growth Regulator Group, Bristol. pp. 57-71.

Jackson, M.B., Davies, D.D. and Lambers, H. (eds) (1990). Plant life under oxygen deprivation. The Hague, The Netherlands, SPB Acad. Publ.

Jackson, M.B., Fenning, T.M., Drew, M.C. and Saker, L.R. (1985). Stimulation of ethylene production and gas space (aerenchyma) formation in adventitious roots of *Zea mays* L. by small partial pressures of oxygen. Planta, **165**: 486-492.

Jackson, W.T. (1960). Effect of IAA on rate of elongation of root hairs of *Agrostis alba* L. Physiologia Plantarum, **13**: 36-45.

Jaunin, F. (1988). Differentiation du rhizoderme chez *Zea mays* L. Thesis, University of Launanne. (cf. Hofer, 1991).

Jaunin, F. and Hofer, R.M. (1986). Root hair formation and elongation of primary maize roots. Physiologia Plantarum **68**: 653-656.

Jaunin, F. and Hofer, R.M. (1988). Calcium and rhizodermal differentiation in primary maize roots. Journal of Experimental Botany, **39**: 587-593.

Jennings, D.H. (1986). Salt relations of cells, tissues, and roots. In: Plant physiology: a treatise vol. IX. Water and solutes in plants (ed. F.C. Steward), pp. 225-379. Academic Press.

Jensen, C.R., Stolzy, L.H. and Letey, J. (1967). Tracer studies of oxygen diffusion through roots of barley, corn and rice. Soil Science, **103**: 23-29.

Jensen, W.A. and Salisbury, F.B. (1984). Botany: an ecological approach. Wadsworth Pub. Co., California.

Johnson-Flanagan, A.M. and Owens, J.N. (1986). Root respiration in white spruce (*Picea glauca*) seedlings in relation to morphology and environment. Plant Physiology (Bethesda), **81**: 21-25.

Jones, H., Tomos, A.D., Leigh, R.A. and Wun Jones, R.G. (1983). Water relations parameters of epidermal and cortical cells in the primary root of *Triticum aestivum* L. Planta, **158**: 230-236.

Justin, S.H.F.W. and W. Armstrong (1991). Evidence for the involvement of ethelene in aerenchyma formation in adventitious roots of rice (*Oryza sativa* L.). New Phytology, **118**: 49-62.

Kapulnik, Y., Okon, Y. and Y. Henis (1985). Changes in root morphology of wheat caused by Azospirillum inoculation. Can. J. Microbiol., **31**: 881-887.

Katagiri, S. (1988). Estimation of the proportion of the root respiration in total soil respiration in deciduous hardwood stands. Journal of Japanese Forestry Society, **70**: 151-158.

Keeley, J.L. (1980). Endomycorrhizae influence growth of Blackgum seedlings in flooded soils. American Journal of Botany, **67**: 6-9.

*Kennedy, R.A. and others (1990). In: 'Plant life under oxygen deprivation' (M.B. Jackson, *et al.*, Eds.), SPB Acad. Publi., The Hague.

Kennedy, R.A., Rumpho, M.E. and Fox, T.C. (1992). Anaerobic metabolism in plants. Plant Physiology, **100**: 1-6.

Khan, A.G. (1974). The occurrence of mycorrhizas in halophytes, hydrophytes and xerophytes and of Endogone spores in adjacent soils. Journal of General Microbiology, **81**: 7-14.

Kiyoshi, K. and Furumoto, M. (1986). A mechanism of respiration-dependent water uptake enhanced by auxin. Protoplasma, **133**: 174-185.

Klepper, B. (1991). Crop root system response to irrigation. Irrigation Science, **12**: 105-108.

Kordan, H.A. (1976). Adventitious root initiation and growth in relation to oxygen supply in germinating rice seedlings. New Phytologist, **76**: 81-86.

Kozai, T., Kubota, C. and Watanabe, I. (1988). Effects of basal medium composition on the growth of carnation plantlets in auto- and mixotrophic tissue culture. Acta Horticulture, **230**: 159-166.

Kozlowski, T.T. (1982). Water supply and tree growth. Part II. Flooding. Forestry Abstr. (CAB) **43**(3): 145-155.

Kozlowski, T.T. (1984a). Responses of woody plants to flooding. In: Flooding and Plant Growth. Ed. T.T. Kozlowski. Acad. Press., New York, pp. 129-163.

Kozlowski, T.T. (1984b). Extent, causes, and impacts of flooding. In: Flooding and Plant growth. T.T. Kozlowski (ed.). Academic Press, New York, pp. 1-6.

Kozlowski, T.T. (1992). Carbohydrate sources and sinks in woody plants. Botanical Review, **58**: 107-222.

Kozlowski, T.T., Kramer, P.J. and Pallardy, S.G. (1991). The physiological ecology of woody plants. Academic Press, San Diego.

Kramer, P.J. (1983). Water relations of plants. Academic Press, New York.

Kramer, P.J. and Boyer, J.S. (1995). Water relations of plants and soils. Academic Press, New York.

Kramer, P.J. and Jackson, W.T. (1954). Causes of injury to flooded tobacco plants. Plant Physiology, **29**, 241-245.

Kramer, P.J. and Kozlowski, T.T. (1979). Physiology of woody plants. Academic Press, N.Y.

Kumar, S.S. and Krishnamurthy, K.V. (1998). The role of root hairs in the mycorrhizal association of the ground orchid, *Spathoglottis plicata* Blume. Intl. J. Parasitol., **28**(9): 15-17.

Kumarasinghe, R.M.K. and Nutman, P.S. (1977). Rhizobium-stimulated callose formation in clover root hairs and its relation to infection. Journal of Experimental Botany, **28**: 961-976.

Kunda Deval, Vasantharajan, V.N. and Bhat, J.V. (1971). Low cation exchange capacity of roots of the root parasite *Santalum album* L. Current Science (Bangalore), **40**(24): 662-663.

Kurkova, E.B. (1981). Distribution of plasmodesmata in root epidermis. In: Structure and function of plant roots (eds. Brouwer *et al.*), pp. 107-109. The Hague.

Laing, H.E. (1940). Respiration of the rhizomes of *Nuphar advenum* and other water plants. American Journal of Botany, **27**: 574-581.

Lakshmi-kumari, M., Singh, C.S. and Subba Rao, N.S. (1974). Root hair infection and nodulation in lucerne (*Medicago sativa* L.) as influenced by salinity and alkalinity. Plant Soil, **40**: 261-268.

Lambers, H. (1987). Growth, respiration, exudation and symbiotic association. The fate of carbon translocated to the roots. In: Root development and function, Eds. Gregory, P.J., Lake, J.V. and Rose, D.A., Cambridge Univ. Press, pp. 125-145.

Lamont, B. (1982). Mechanisms for enhancing nutrient uptake in plant, with particular reference to Mediterranean South Africa and Australia. Botanical Review, **48**: 597-689.

Layzell, D.B. and Hunt, S. (1990). Oxygen and the regulation of nitrogen fixation in legume nodules. Physiol. Plant., **80**: 322-327.

Lee, B. and Priestley, J.H. (1924). Annals of Botany, **38**: 525.

Lefebvre, D.D. (1985). Stomata on the primary root of *Pisum sativum* L. Annals of Botany, **55**: 337-341.

Lemon, E.R. (1962). Soil aeration and plant root relations. II. Root respiration. Agronomic Journal, **54** : 171-175.

Leonard, O.A. and Pinckard, J.P. (1946). Effect of various oxygen and carbon dioxide concentrations on cotton root development. Plant Physiology, **21**: 18-36.

Letey, J., L.H. Stolzy and G.B. Blank (1962). Effect of duration and timing of low soil oxygen content on shoot and root growth. Agronomic Journal, **54**: 34-37.

Levitt, J. (1953). Further remarks on the thermodynamics of active non-osmotic water absorption. Physiologia Plantarum, **6**: 240-252.

Levitt, J. (1974). Introduction to Plant Physiology. 2nd ed. The C.V. Mosby Co., St. Louis.

Lew, R.R. (1991). Electrogenic transport properties of growing Arabidopsis root hairs: the plasma membrane and potassium channels. Plant Physiology, **97**: 1527-1534.

Lewis, D.G. and Quirk, J.P. (1967). Phosphate diffusion in soil and uptake by plants. III. P^{31} movement and uptake by plants as indicated by P^{32} autoradiography. Plant & Soil, **26**: 445-453.

Lindsay, J.I., Osci-Yeboah, S. and Gumbs, F.A. (1983). Effects of different tillage methods on maize growth on a tropical Inceptisol with impeded drainage. Soil Tillage Research, **3**: 185-196.

Lotocka, B., Kooinska, J., Gorecka, M. and Golinowski, W. (2000). Formation and absorption of root nodule primordial in *Lupinus luteus* L. Acta Biologica Cracoviensia. Ser. Botanica, **42**(1): 87-102.

Ludwig, F. (1881). Ueber die Bestaubungverhaltnisse einiger Susswasserpflanzen und ihre Anpassungen an das Wasser und gewisse wasserbewohnende Insekten. Kosmos (Stuttgart). Jahrg. **10**: 7-12. (cf. Shannon, 1953).

Lundegardh, H. (1955). Mechanisms of absorption, transport, accumulation, and secretion of ions. Annual Review of Plant Physiology, **6**: 1-24.

Luxmoore, R.J. and L.H. Stolzy (1972). Oxygen diffusion in the soil-plant system. VI. A synopsis with commentary. Agron. J. **64**: 725-729.

MacDonald, R.C. and T.W. Kimmerer (1991). Ethanol in the stems of trees. Physiologia Plantarum, **82**: 582-588.

Machilis, L. (1944). The respiratory gradient in barley roots. American Journal of Botany, **31**: 281-282.

Mackay, A.D. and Barber, S.A. (1984). Comparison of root and root hair growth in solution and soil culture. Plant Nutrition, **7**: 1745-1757.

Mackay, A.D. and Barber, S.A. (1985). Effect of soil moisture and phosphate level on root hair growth of corn roots. Plant & Soil, **86**: 321-331.

Mackay, A.K. and Barber, S.A. (1987). Effect of cyclic wetting and drying of a soil on root hair growth of maize roots. Plant & Soil, **104**: 291-293.

Maeda, M. (1954). The meaning of mycorrhiza in regard to systematic botany. Kuamoto Journal Science, Sect. B. **3**: 57-84.

Mamaev, V.V. (1984). Respiration of tree roots in Pinetum and Betuleum oxalidoso myrtillosum. Lesouvedemie, **6** : 53-60.

Manjunath, A., Mohan, R., Raj, J. and Bagyaraj, D.J. (1981). Vesicular-arbuscular mycorrhizas in cultivars of rice. Journal Soil Biology & Ecology, **1**: 1-4.

Manna, P.S., Sen, S., Ghosh, S.K. and Jana, T.K. (1992). Study of diffusion of O_2 between atmosphere and surface water in the Hooghly estuary. Tropical Ecology, **33**: 186-190.

Martin, J.T. and Juniper, B.E. (1970). The cuticles of plants. St. Martin's Press, NY.

Mascarenhas, A.F., Hendre, R.R., Nadgir, A.L., Barne, D.M. and Jagannathan, V. (1978). Differentiation in tissue culture of cabbage. Indian Journal Experimental Biology, **16**: 122-125.

McClelland, M.T. and Smith, M.A.L. (1988). Response of woody plant micro-cuttings to *in vitro* and *ex vitro* rooting methods. Combined Proceedings International Plant Propagators' Society, **38**: 593-599.

McCully, M.E. (1987). Selected aspects of the structure and development of field-grown roots with special reference to maize. In: Root development and function, Gregory *et al.* (Eds.), Cambridge Univ. Press, pp. 53-70.

McDougall, W.B. (1921). Thick-walled root hairs of *Gleditsia* and related genera. American Journal of Botany, **8**: 171-175.

McPherson, D.C. (1939). Cortical air spaces in the roots of *Zea mays* L. New Phytologist, **38**: 190-202.

Meisner, C.A. and Karnok, K.J. (1991). Root hair occurrence and variation with environment. Agronomic Journal (USA), **83**(5): 814-818.

Mejstrik, V. (1976). The ecology of mycorrhiza in plants from peat bog areas of the Trebon Basin in relation to the ground water table. Quest. Geobiol., **16**: 99-174.

Mer, E. (1879). Recherches experimentales sur les conditions de developement des poils radicans. Compt. Red. Acad. Sci. Paris, **88**: 665-668.

Mexal, J.G., Reid, C.P.P. and Burke, E.J. (1979). Scanning electron microscopy of lodgepole pine roots. Botanical Gazette, **140**: 318-323.

*Meyen, E.J.E. (1838). Neves system der pflanzen physiologie. Bd. **2**: 16, 19.

Michelsen, A. (1993). The mycorrhizal status of vascular epiphytes in Bale Mountains National Park, Ethiopia. Mycorrhiza, **4** (1) : 11-15.

Miller, I.M. and Baker, D. (1986). Nodulation of actinorhizal plants by Frankia strains capable of both root infection and intercellular penetration. Protoplasma **131**(1): 82-91.

*Molisch, H.(1885). Uber die Ablenkung der wurzeln von ihrer normalen wachstumsrichtung durch gase (aerotropismus). Sitzungsber. Akad. Wiss. Wien, **90**: 111-196.

Monk, L.S. and Brandle, R. (1982). Adaptation of respiration and fermentation to changing levels of oxygen in rhizomes of *Schoenoplectus lacustris* (L.) Palla and its significance to flooding tolerance. Zeitschrift fur Pflanzenphysiologie, **105**: 369-374.

Mori, S. and Hagihara, A. (1991). Root respiration in *Chamaecyparis obtusa* trees. Tree Physiology, **8**: 217-225.

Mort, A.J. and Grover, P.B. Jr. (1988). Characterization of root hair cell walls as potential barriers to the infection of plants by Rhizobia. Plant Physiology, **86**: 638-641.

Munns, D.N. (1968). Nodulation of *Medicago sativa* in solution culture. I. Acid sensitive steps. Plant Soil, **28**: 129-146.

Murray, L.E. and Christianson, M.L. (1987). Phylogenetic comparison of large nuclear DNA contents of differentiating cells in the roots of *Equisetum, Tradescantia* and *Hordeum*. American Journal of Botany, **74**: 1772-1778.

Murray, L.E., Christianson, M.L., Alfinito, S.H. and Garger, S.J. (1987). Characterization of the nuclear DNA of *Hordeum vulgare* root hairs: amplification disappears under salt stress. American Journal of Botany, **74**: 1779-1786.

Nambiar, E.K. (1976). Uptake of Zn^{65} from dry soil by plants. Plant & Soil, **44**: 267-271.

Nambiar, P.T.C., Nigam, S.N., Dart, P.J. and Gibbons, R.W. (1983). Absence of root hairs in non-nodulating groundnut (*Arachis hypogea*). Acta Botanica Neerlandica, 32: 29-38 and Journal of Experimental Botany, **34**: 484-488.

Nanda, R. and Saini, A.D. (1992). Effect of restricted soil moisture on yield and its attributes in chickpea. Indian Journal of Plant Physiology, **35**: 16-24.

Negbi, M. and Koller, D. (1962). Homologies in the grass embryo—a re-evaluation. Phytomorphology, **12**: 289-296.

Newcomb, E.H. and Bonnett, H.T. (1965). Cytoplasmic microtubule and wall microfibril orientation in root hairs of radish. Journal of Cell Biology, **27**: 575.

Newman, E.I. (1974). Root and soil water relationships. In: The plant root and its environment (ed. by E.W. Carson), University Press of Virginia.

Newman, E.I. and Reddell, P. (1987). The distribution of mycorrhizas among families of vascular plants. New Phytol. **106**: 745-751.

Nieuwdorp, P.J. (1972). Some observations with light and electron microscope on the endotrophic mycorrhiza of orchids. Acta Bot. Neerl., **21**: 128-144.

Nirmal Ram, Sheelwant Patel and Gupta, M.P. (1993). Effect of compactness by timber operations on the infiltration rate in a sal plantation. Indian Forester, **119**: 173-179.

Noda, T., Tanaka, N., Mano, Y., Nabeshima, S., Ohkawa, H. and Matsui, C. (1987). Regeneration of horse radish hairy roots incited by *Agrobacterium rhizogenes* infection. Plant Cell Reports, **6**: 283-286.

Noll, F. (1903). In Lehrbuch der Botanik fur Hochschulen by Strasburger, E., Fritz Noll, H. Schenck and G. Karsten. pp. 197-213.

Norris, F. de la M. (1913). Production of air passage in the root of *Zea mays* by variation of culture media. Proc. Bristol Nat. Soc., **4**: 134-138.

North, G.B. and Nobel, P.S. (1992). Drought-induced changes in hydraulic conductivity and structure in roots of *Ferocactus acenthodes* and *Opuntia ficus-indica*. New Phytologist, **120**(1): 9-19.

Ohlendorf, H. (1985). Symbiosis of legume with *Rhizobium* band mechanisms controlling the infection. Angew Bot., **59** :279-303.

Ohlert, E. (1837). Einige bemergungen uber die wurzelzasern der hoheren pflanzen. Linnaea, **1837** : 609, 624, 628.

Okon, Y., Fallik, E., Yahalom, E. and Tal, S. (1988). Plant growth promoting effects of *Azospirillum*. In: Nitrogen fixation: Hundred years after (Botha, H. *et al.* Eds.). Gustav Fischer, Stuttgart, West Germany, pp. 741-746.

Oliveira, L. (1977). Changes in the ultra-structure of mitochondria of roots of *Triticale* subjected to anaerobiosis. Protoplasma, **91**: 267-280.

Opik, H. (1973). Effect of anaerobiosis on respiratory rate, cytochrome oxidase activity and mitochondrial structures in coleoptiles of rice (*Oryza sativa* L.). Journal of Cell Science, **12:** 725-739.

Orion, D. and Lapid, D. (1993). Scanning electron microscope study on the interactions of *Pratylenchus mediterraneus* and *Vicia sativa* roots. Nematologica, **39**(3): 322-327.

Palladin, V.I. 1918. Plant Physiology. Blakiston's Son and Co. (cf. Whitaker, 1923).

Parthasarathi, K., Gupta, S.K. and Rao, P.S. (1974). Differential response in the cation exchange capacity of the host plants on parasitization by sandal (*Santalum album* L.). Current Science, **43**(1): 20.

Patriquin, D.G., Dobereiner, J. and Jain, D.K. (1983). Sites and processes of association between diazotrophs and grasses. Canadian Journal of Microbiology, **29**: 900-915.

Pezeshki, S.R. (1991). Root responses of flood tolerant and flood sensitive species to soil redox. Trees, **5**(3): 180-186.

Peterson, R.L. and Farquhar, M.C. (1996). Root hairs: specialized tubular cells extending root surfaces. The Bot. Rev., **62**(1): 2-17.

Pfeffer, W. (1897). Planzen Physiologie I. Leipzig (cf. E.S. Whitaker 1923). Root hairs and secondary thickening in the Compositae. Botanical Gazette, **76**: 30-59.

Philipson, J. J. and Coutts, M.P. (1980). The tolerance of tree roots to waterlogging. New Phytol., **85**: 489- 494.

Phillipson, J., Putman, R.J., Steel, J. and Woodall, S.R.J. (1975). Litter input, litter decomposition and the evolution of carbon dioxide in a beech woodland, Wytham Woods, Oxford. Oecologia, **20**: 203-217.

Pill, W.G. and Beers, E.P. (1987). Effect of germination in humid air and aerated water on root hair frequency and emergence of fluid-drilled tomato seeds. Applied Agricultural Research, **1**: 294-296.

Pitts, R.J., Cernac, A. and Estelle, M. (1998). Auxin and ethylene promote root hair elongation in *Arabidopsis*. Plant Journ., **16**(5): 553-560.

Poerwanto, R., Hiroshi, I. and Ikuo Kataoka (1987). Observations on the root hairs of three citrus root stocks. Kagawa Daigaku Nogakubu Gakujutsu Hokoku, **39**: 5-10 (En. BA 90155/1988).

Popham, R.A. and Henry, R.D. (1955). Multicellular root hairs on adventitious roots of *Kalanchoe fedtschenkoi*. Ohio Journal of Science, **55**: 301-307.

Pridgeon, A.M. (1987). The valamen and exodermis of orchid roots. In: Arditti, J. (ed.) Orchid biology, reviews and perspectives, IV. Cornell Univ. Press, Ithaca (cf. Romberger *et al.*, 1993).

Prinsloo, G.C., Baard, S.W. and Ferreira, J.F. (1992). A scanning electron microscope study of the infection and colonisation of chicory roots by *Thielaviopsis basicola* Phytophylactica, **24**(3): 293-296.

Rama Rao, (1903). Root parasitism of Sandal tree. Indian Forester, **29**: 386.

Ranson, S.L. and Parija, B. (1955). Experiments on growth in length of plant organs. Journal of Experimental Botany, **6**: 80-93.

Rasheed, J.H., AlMallah, M.K., Cocking, E.C. and Davey, M.R. (1990). Root hair protoplasts of *Lotus corniculatus* L. (birdsfoot trefoil) express their totipotency. Plant Cell Reports, **8**: 565-569.

Raskin, I.K.H. (1985). Mechanism of aeration in rice. Science, **228**: 327-329.

Raven, J.A., Handley, L.L., Macfarlane, J.J., McInroy, S., McKenzie, L., Richards J.H., Samauelsson, G. (1988). The role of CO_2 uptake by roots and CAM in acquisition of inorganic C by plants of the isoetid life-form: a review, with new data on *Eriocaulon decangulare* L. New Phytol., **108**: 125-148.

Rawson, H.M. and Turner, N.C. (1982). Recovery from water stress in five sunflower (*Helianthus annuus* L.) cultivars—II. The development of leaf area. Australian Journal of Plant Physiology, **9**: 449-460.

Read, D.J., Kouchiki, H.K. and Hodgson, J. (1976). Vesicular-arbuscular mycorrhiza in natural vegetation systems. New Phytologist, **77**: 641-653.

Reeder, J.R. and von Maltzahn, K. (1953). Taxonomic significance of root-hair development in the Gramineae. Proceedings of National Academic Sciences (US), **39** : 593-598

Reid, (1987). The role of ethylene. In: Plant hormones and their role in plant growth and development, Ed. Davies, P.J. Martinus-Nijhoff Publ.

Reid, C.P.P. and Bowen, G.D. (1979). Effects of soil moisture on V/A mycorrhiza formation and root development in Medicago. In: The soil-root interface, Eds. J.L. Harley and R.S. Russel. Academic Press, London.

Reiss, H.D. and Herth, W. (1979). Calcium gradients in tip growing plant cells visualized by chlorotetracycline fluorescence. Planta, **146**: 615-621.

Reynolds, E.R.C. (1975). Tree rootlets and their distribution. In: The development and function of roots, Eds. Torrey, J.G. and D.T. Clarkson. Academic Press, London.

Richardson, S.D. (1953). A note on some differences in root hair formation between seedlings of Sycamore and American oak. New Phytologist, **52**: 80-82.

Ridge, R.W. (1988). Freeze-substitution improves the ultrastructural preservation of legume root hairs. Botanical Magazine (Tokyo), **101**: 427-441.

Robinson, D. and Rorison, I.H. (1987). Root hairs and plant growth at low nitrogen availabilities. New Phytologist, **107**: 681-193.

Romberger, J.A., Z. Hejnowicz and J.F. Hill (1993). Plant structure: Function and development—a treatise on anatomy and vegetative development, with special reference to woody plants. Springer-Verlag, Berlin.

Row, H.C. and Reeder, J.R. (1957). Root-hair development as evidence of relationships among genera of Gramineae. American Journal of Botany, **53**: 44:596-601.

Rosen, C.J. and R.M. Carlson (1984). Influence of root zone oxygen stress on potassium and ammonium absorption by Myrobalan plum rootstock. Plant & Soil, **80**: 345-353.

Rosene, H.F. (1943). Quantitative measurement of the velocity of water absorption in individual root hairs by microtechnique. Plant Physiology, **18**: 588-608.

Rosene, H.F. (1950) Ageing and the influx of water into radish root hair cells. Journal of General Physiology, **34**: 65.

Rosene, H.F. (1954). A comparative study of the rates of water influx into the hairless epidermal surface and the root hairs of onion roots. Physiologia Plantarum, 7 : 676-686.

Rosene, H.F. and Walthall, A.M.J. (1949). Velocities of water absorption by individual root hair of different species. Botanical Gazette, **111**: 11-21.

Rosene, H.F. and Walthall, A.M.J. (1954). Comparison of the velocities of water influx into young and old root hairs of wheat seedlings. Physiologia Plantarum, **7**: 190-194.

Rovira, A.D. and Bowen, G.D. (1968). Anion uptake by the apical region of seminal wheat roots. Nature (London), **218**: 685-686.

Row, H.C. and Reeder, J.R. (1957). Root hair development as evidence of relationships among genera of Gramineae. American Journal of Botany, **44**: 596-601.

Russel, R.S. (1977). Plant root systems: their function and interaction with the soil. McGraw-Hill Book Co., UK.

Russo, R.O. and Berlyn, G.P. (1990). The use of organic biostimulanats to help low input sustainable agriculture. Journal of Sustainable Agriculture, **1** : 19-42.

*Sachs (1990). In: Plant life under oxygen deprivation. (M.B. Jackson *et al.*, Eds.), SPB Acad. Publ., The Hague. (Cf. Jackson *et al.*, 1990)

Safir, G.R., Boyer, J.S., Gerdemann, J.W. (1971). Mycorrhizal enhancement of water transport in soybean. Science, **172**: 581-583.

Saglio, P.H., Raymond, P. and Pradet, A. (1983). Oxygen transport and root respiration of maize seedlings. A quantitative approach using the correlation between ATP/ADP and the respiration rate controlled by oxygen tension. Plant Physiology, **72**: 1035-39.

Said, A.G.E. and Murashige, T. (1979). Continuous cultures of tomato and citrus roots in vitro. *In Vitro*, **15**: 593-602. (cf. George, 1993).

Saksena, R. S. (1992). Water management under water sprinkler and drip irrigation. In: Irrigation Management in Agriculture, Vol. II, pp. 637-667. WWF, New Delhi.

Samantarai, B. (1938). Respiration of amphibious plants. I. Journal of Indian Botanical Society, **17**: 195-204.

Sand-Jensen, K., Prahl, C. and Stokholm, H. (1982). Oxygen release from roots of submerged aquatic macrophytes. *Oikos*, **38**: 349-354.

Sasson, M.A., Wolters, A.M.C. and Traas, J.A. (1985). Deposition of cellulose micro-fibrils in cell walls of root hairs. European Journal of Cell Biology, **37**: 21-26.

Saur, E., Bonheme, I., Nygren, P. and Imbert, D. (1998). Nodulation of *Pterocarpus officinalis* in the swamp forest of Guadeloupe (Lesser Antilles). J. Trop. Ecol., **14**(6): 761-770.

Schiefelbein, J.W. and Somerville, C. (1990). Genetic control of root hair development in *Arabidopsis thaliana*. The Plant Cell (USA), **2**: 235-243.

Schiefelbein, Joh W., Shipley, A. and Paul Rowse (1992). Calcium influx at the tip of growing root hair cells of *Arabidopsis thaliana*. Planta (Heidelb), **187**: 455-459.

Scholander, P.F., van Dam, L. and Sholander, S.I. (1955). Gas exchange in the roots of mangroves. American Journal of Botany, **42**: 92-98.

Scott, F.M. (1963). Nature (London) 199: 1009. (cf. Martin and Juniper, 1970).

Scott, F.M. (1966). Nature (London) 210: 1015. (cf. Martin and Juniper, 1970).

Sculthorpe, C.D. (1967). The biology of aquatic vascular plants. London: Edward Arnold.

Sculthorpe, C.D. (1985). The biology of aquatic vascular plants. Koeltz Sci. Books, West Germany.

Seagull, R.W. and Heath, L.B. (1980). The organization of cortical microtubule arrays in the radish root hair. Protoplasma, **103** : 205-229.

Sena Gomes, A.R. and Kozlowski, T.T. (1980a). Growth responses and adaptations of *Fraxinus pennsylvanica* seedlings to flooding. Plant Physiology, **66**: 267-271.

Sena Gomes, A.R. and Kozlowski, T.T. (1980b). Responses of *Melaleuca quinquenervia* seedlings to flooding. Physiologia Plantarum, **49**: 373-377.

Sena Gomes, A.R. and Kozlowski, T.T. (1980c). Effects of flooding on growth of *Eucalyptus camaldulensis* and *E. globulus* seedlings. Oecologia, **46**: 139-142.

Shanmughan, C.R. (1992). Concepts and approaches in agricultural irrigation. In: Irrigation Management in Agriculture, Vol. I, pp. 263-274, WWF, New Delhi.

Shannon, E.C. (1953). The production of root hairs by aquatic plants. American Midland Naturalist, **50**: 474-479.

Shaw, S.L., Dumais, J. and Long, S.R. (2000). Cell surface expansion in polarly growing root hairs of *Medicago truncatula*. Plant Physiology, **124**(3): 959-969.

Shuja, N., Gilani, U. and Khan, A.G. (1971). Mycorrhizal associations in some angiosperm trees around New University Campus, Lahore. Pakistan Journal Forestry, **24** : 367-374.

Sieberer, B. and Emons, A.M.C. (2000). Cytoarchitecture and pattern of cytoplasmic streaming in root hairs of *Medicago truncatula* during development and deformation by nodulation factors. Protoplasma, **214**(1-2): 118-127.

Sierp, H. and Brewig, A. (1935). Quantitative untersuchungen uber die wasserabsorptionzone des wurzeln. Jahrbucher fur Wissenschaftliche Botanik, **82**: 99-122.

Sievers, A. (1963a). Beteiligung des Golgi apparatus beider bildung der Zellwand von Wurzelhaaren. Protoplasma, **56**: 188-192.

Sievers, A. (1963b). Uber die feinstruktur des plasmas wachsender wurzelhaare. Zeitschrift fur Naturforschung, **186**: 830-836.

Sievers, A. and Schnepf, E. (1981). Morphogenesis and polarity of tubular cells with tip growth. In: Cytomorphogenesis in plants, ed. Kiermayer, O. Springer-Verlag, New York, pp. 265-299.

Simpson, L.A. and Gumbs, F.A. (1992). Effect of continued cropping on a heavy clay soil on the coast of Guyana with and without tillage. Tropical Agriculture (Trinidad), **69**: 111-118.

Smit, G., Swart, S., Lugtenberg, B.J.J. and Kijne, J.W. (1992). Molecular mechanisms of attachment of *Rhizobium* bacteria to plant roots. Molecular Microbiology, **6** (20): 2897-2903.

Smith, S.E. and Gianinazii-Pearson (1988). Physiological interactions between symbionts in vesicular-arbuscular mycorrhizal plants. Ann. Rev. Pl. Physiol. Mol. Biol., **39**: 221-244.

Sondergaard, M. and Laegaard, S. (1977). Vesicular-arbuscular mycorrhiza in some aquatic vascular plants. Nature (London), **268**: 232-233.

Spaink, H.P. (1992). Rhizobial lipo-oligosaccharides: answers and questions. Plant Mol. Biol., **20**: 977-986.

Sprent, J.I. (1972). The effect of water stress on nitrogen-fixing root nodules II. Effects on the fine structure of detached soybean nodules. New Phytologist, **71**: 443.

Steinmann, F. and R. Brandle (1981). Die uberflutungstoleranz der teichbinse (*Schoenoplectus lacustris* (L.) Palla): III. Beziehungen zwischen der Sauerstroffversorgung und der 'Adenylate Energy Charge' der rhizome in Abhangigkeit von der sauerstoffkonzentration in der umgebung. Flora, **171**: 307-314.

St. John, T.V. (1980). Root size, root hairs and mycorrhizal infection: A reexamination of Baylis's hypothesis with tropical trees. New Phytologist, **84**: 483-487.

Stolzy, L.H. (1972). Soil aeration and gas exchange in relation to grasses. In: Younger, U.B., Mitchell, C.M. eds. Biology and utilization of grasses. Academic Press, N.Y., pp. 247-258.

Studer, C. and Brandle, R. (1984). Sauerstoffkonsum und Versorgung der Rhizome von *Acorus calamus* L., *Glyceria maxima* (Hartmann) Holmberg, *Menyanthes trifoliata* L., *Phalaris arundinacea* L., *Phragmites communis* Trin. und *Typha latifolia* L. Botanica Helvetica **94**: 23-32.

Sumio, I. and Barber, S.A. (1983). Phosphate uptake by 6 plant species as related to root hairs. Agronomic Journal, **75**: 457-461.

Sun, C.N. (1955). Growth and development of primary tissues in aerated and non-aerated roots of soybean. Bulletin of Torrey Botanic Club, **82**: 491-502.

Talbot, R.J., Etherington, J.R. and Bryant, J.A. (1987). Comparative studies of plant growth and distribution in relation to waterlogging. XII. Growth, photosynthetic capacity and metal ion uptake in *Salix caprea* and *S. cinerea* sp. *oleifolia*. New Phytologist, **105**: 563-574.

Tanaka, Y. and Woods, F.W. (1972). Root and root hair growth in relation to supply and internal mobility of calcium. Botanical Gazette, **133**: 29-34.

Tang, C.X., Kuo, J., Longnecker, N.E., Thomson, C.J. and Robson, A.D. (1993). High pH causes disintegration of the root surface in *Lupinus angustifolius* L. Annals of Botany, **71**(3): 201-207.

Tarkowska, J.A. and Wacowska, M. (1988). The significance of the presence of stomata on seedling roots. Annals of Botany, **61**: 305-310.

Thumfort, P.P., Atkins, C.A. and Layzell, D.B. (1994). A re-evaluation of the role of the infected cell in the control of O_2 diffusion in legume nodules. Plant Physiol., **105**: 1321-1333.

Tietz, A. and Urbasch, I. (1977). Spaltoffnungen an der keimwurzel von *Helianthus annuus* L. Naturwissenschaften, **64**: 10.

Tinker, P.B. (1976). Roots and water. Transport of water to plant roots in soil. Philadelphia Transaction of Royal Society (London), Series B, **273**: 445.

Topa, M.A. and McLeod, K.W. (1986). Aerenchyma and lenticel formation in pine seedlings: a possible avoidance mechanism to anaerobic growth conditions. Physiologia Plantarum, **68**: 540-550.

Traas, J.A., Braat, P., Emons, A.M.C., Meekes, H., Derksen, J. (1985). Microtubules in root hairs. Journal of Cell Science, **76**: 303-320.

Trappe, J.M. (1987). Phylogenitic and ecologic aspects of mycotrophy in the angiosperms from an evolutionary standpoint. In: Ecophysiology of VA mycorrhizal plants (G.R. Safir, Ed.). CRC Press, Boca Raton, Fla., pp. 5-25.

*Troll, W. (1948). Allgemeine Botanik. F. Enke, Stuttgart.

Trought, M.C.T. and Drew, M.C. (1980). The development of water logging damage in young wheat plants in anaerobic solution cultures. Journal of Experimental Botany, **31**: 1573-85.

Turner, N. C. (1990). Plant water relations and irrigation management. Agricultural Water Management, **17**: 59-73.

Tyree, M.T. (1970). The symplast concept: a general theory of symplast transport according to the thermo-dynamics of irreversible processes. Journal of Theoretical Biology, **26** : 181-124.

Umali-Garcia, M., Hubbell, D.H., Gaskins, M.H. and Dazzo, F.B. (1980). Association of *Azospirillum* with grass roots. Applied Environmental Microbiogy, **39**: 219-226.

Uphof, J.C.T.H. (1962). Plant hairs. In Handbuch der Pflanzenanatomie, 2nd ed., Vol. IV, Part 5. Borntraeger, Berlin, pp. 1-206.

Usha M., Asplund, P.T., Curtis, W.R. and Mukundan, U. (2000). Cytochemical assay to localize respiratory activity in hairy root cultures. Phytomorphology, **50**(2): 184-187.

*Vakhimistrov, D.B. and Zlotnikova, I.F. (1990). Functional specificity of root hairs. Fiziologiya-Rastenii, **37**: 936-954.

van Noordwijk, M. Brouwer, G. and Harmanny, K. (1993). Concepts and methods for studying interactions of roots and soil structure. Geoderma, **56**: 351-375.

van Noordwijk, M. and de Willigen, P. (1984). Mathematical models on diffusion of oxygen to and within plant roots, with special emphasis on effects of soil-root contact. II. Application. Plant & Soil, **77**: 233-241.

Varadaraja Iyengar, A.V. (1965). The physiology of root-parasitism in sandal (*Santalum album* Linn.). Indian Forester, **91**: 246-258.

Vartanian, N., Wertheimer, D.S. and Couderc, H. (1983). Scanning electron microscopic aspects of short tuberized roots, with special reference to cell rhizodermis evolution under drought and rehydration. Plant Cell Environment, **6**: 39-46.

*Vartapetyan, B.B. (1973). Aeration of roots in relation to molecular oxygen transport in plants. In: Plant response to climatic factors. Proc. Uppsala Symp. 1970 (ecology and conservation, 5) pp. 259-265. United Nations Educational Scienitific and Cultural Organization (cf. Jackson *et al.*, 1985).

Vartapetyan, B.B. (1983). Anaerobiosis and the theory of physiological adaptation of plants to flooding. Soviet Plant Physiology, **29**: 764-771.

Vartapetyan, B.B. (1990). In: Plant life under oxygen deprivation (M.B. Jackson *et al.*, Eds.), SPB Acad. Publ., The Hague.

Vartapetyan, B.B. and Jackson, M.B. (1997). Plant adaptations to anaerobic stress. Ann. Bot. **79** (Suppl. A.): 3-20.

Vartapetyan, B.B., Andreeva, I.N., Kozlova, G.I. and Agapova, L.P. (1977). Mitochondrial ultra structure in roots of mesophyte and hydrophyte at anoxia and after glucose feeding. Protoplasma, **91**: 243-256.

Vartapetyan, B.B., Andreeva, I.N. and Nuritdinov, N. (1978). Plant cells under oxygen stress. In: Plant life in anaerobic environments (eds D.D. Hook and R.M.M. Crawford), pp. 13-88. Ann Arbor Science Press, Michigan.

Volkmann, D. (1984). The plasma membrane of growing root hairs is composed of zones of local differentiation. Planta (Berlin), **162**: 392-403.

von Guttenberg, H. (1968). Der primare Bau der angiospermenwurzel. In: Handbuch der pflanzenanatomie, 2nd ed. Bornstraeger, Berlin.

Voorrips, R.E. (1992). Root hair infection by *Plasmodiophora brassicae* in club root-resistant and susceptible genotypes of *Brassica oleracea*, *B. rapa* and *B. napus*. Netherlands Journal of Plant Pathology, **98** (6): 361-368.

Weisenseel, M.H., Dorn, A. and Jaffe, L.F. (1979). Natural H$^+$ currents traverse growing roots and root hairs of barley (*Hordeum vulgare* L.). Plant Physiology, **64**: 512-518.

Werner, Dietrich and Adres Bernd Wolff (1987). Root hair specific protein in *Glycine max*. Z. Naturforsch Sect. C. Biosci., **42** : 537-541.

Whitaker, E.S. (1923). Root hairs and secondary thickening in the Compositae. Botanical Gazette, **76**: 30-59.

White, K.J. (1991). Teak: some aspects of research and development. RAPA Publication, 1991/17.

Wiedenroth, E.M. and Erdmann, B. (1985). Morphological changes in wheat seedlings (*Triticum aestivum* L.) following root anaerobiosis and partial pruning of the root system. Annals of Botany, **56**: 507-16.

Wiedenroth, E.M. and Jackson, M.B. (1993). Depletion of oxygen in roots of wheat and maize by respiration following encasement in olive oil. Annals of Botany, **71**: 427-430.

de Willigen, P. and van Noordwijk, M. (1984). Mathematical models on diffusion of oxygen to and within plant roots with special emphasis on effects of soil-root contact. I. Derivation of the models. Plant & Soil, **77**: 215-231.

Wood, S.M. and Newcomb, W. (1989). Nodule morphogenesis: the early infection of alfalfa (*Medicago sativa*) root hairs by *Rhizobium meliloti*. Canadian Journal of Botany, **67**: 3108-3122.

Worall, V.S. and Roughley, R.J. (1976). The effect of moisture stress on infection of *Trifolium subterraneum* L. by *Rhizobium trifolii* Dang. Journal of Experimental Botany, **27**: 1233-1241.

Wulfsohn, D. and Nyengaard, J.R. (1999). Simple stereological procedure to estimate the number and dimensions of root hairs. Plant and Soil, **209**(1): 129-136.

Wymer, C.L., Bibikova, T.N. and Gilroy, S. (1997). Cytoplasmic free calcium distributions during the development of root hairs of *Arabidopsis thaliana*. Plant J., **12**(2): 427-439.

Yang, W.M., Jeong, Y.G., Kang, Y.G. and Cho, J.Y. (1991). The basic study of a new soilless culture system. I. The effects of oxygen concentration of nutrient solution on the physio-ecological and morphological characteristics of tomato. Journal of Korean Society of Horticultural Science, **32**(3): 305-313.

Ycas, J.W. and Zobel, R.W. (1983). The response of maize radicle orientation to soil solution and soil atmosphere. Plant & Soil, **70**: 27-35.

Yemm, E.W. (1965). The respiration of plants and their organs. In: Steward, F.C. (Ed.) Plant Physiology: A Treatise. Vol. IV-A. Acad. Press, London. pp. 231-310.

Yie, S.T. and Liah, S.I. (1977). *In vitro*. **13**: 564-568.

Yoav Waisel and Moshe Agami (1991). Ecophysiology of roots of submerged aquatic plants. In: Plant roots—The hidden half (Eds. Youav Waisel *et al.*). Marcel Dekker Inc., New York.

Young, J.A. and E. Martens (1991). Importance of hypocotyl hairs in germination of *Artemisia* seeds. Journal of Range Management, **44:** 438-442.

Yu, P.T., Stolzy, L.H. and Letey, J. (1969). Survival of plants under prolonged flooding. Agronomy Journal, **61**: 844-847.

*Zlotnikova,I.F. and Vakhmistrov, D.B. (1989). Segregation of cation and anion uptake in root cells. Dokladu Bot. Sci., **307-309**: 63-65.

---

*Original not seen

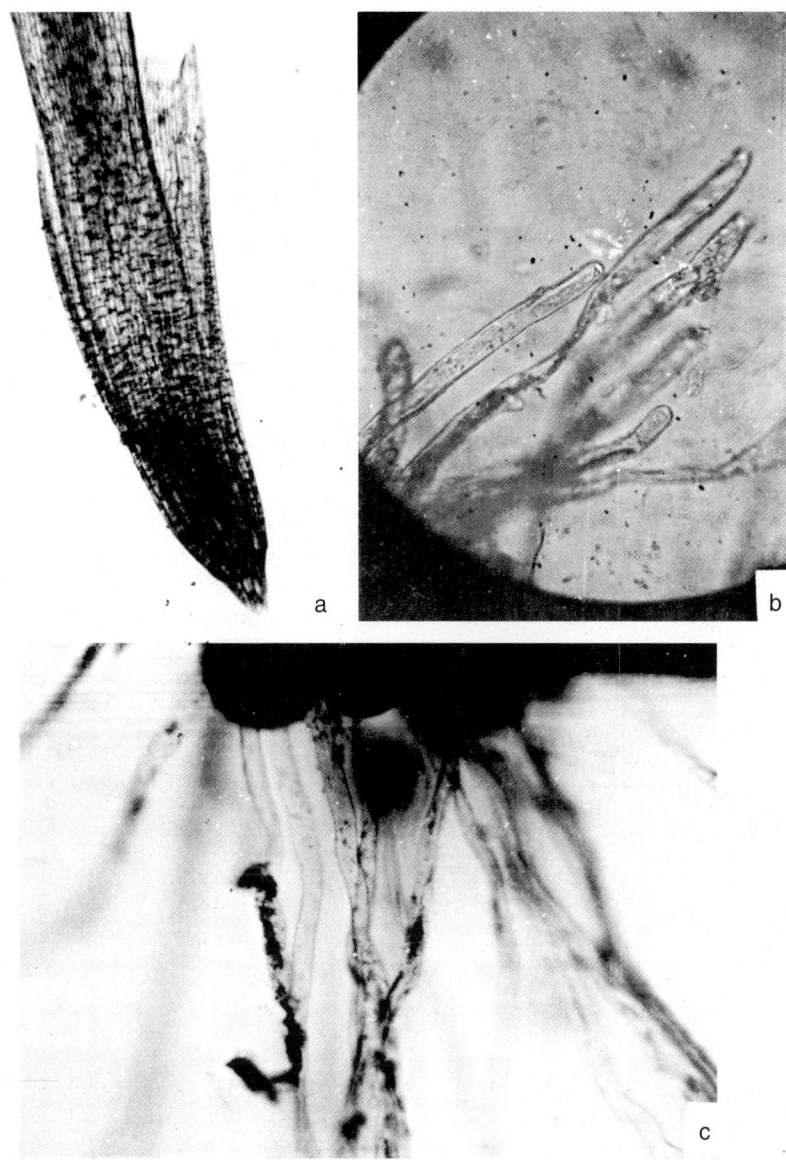

Fig. 1

a. A multi cellular root lateral in *Eichhornia crassipes* which was reported as root hairs by Bagyaraj et al., 1979. Root hairs are absent in this plant (x 375)

b. "Tip callose" in the mature root hairs of *Syzygium cumini* (x 420)

c. Unseptate root hairs in *Arachis hypogea* (Chandler, 1978, reported that root hairs in this crop are septate) (x 350)

Fig. 2

a. Root hairs of exposed radicle in *Ocimum gratissimum* (x 560)

b. Dense root hairs evenly distributed throughout the length of slender radicle in *Casuarina equisetifolia* (x 120)

c. Vestigial root hairs on the radicle of an young recruit in the root parasite, *Santalum album* (x 42.5)

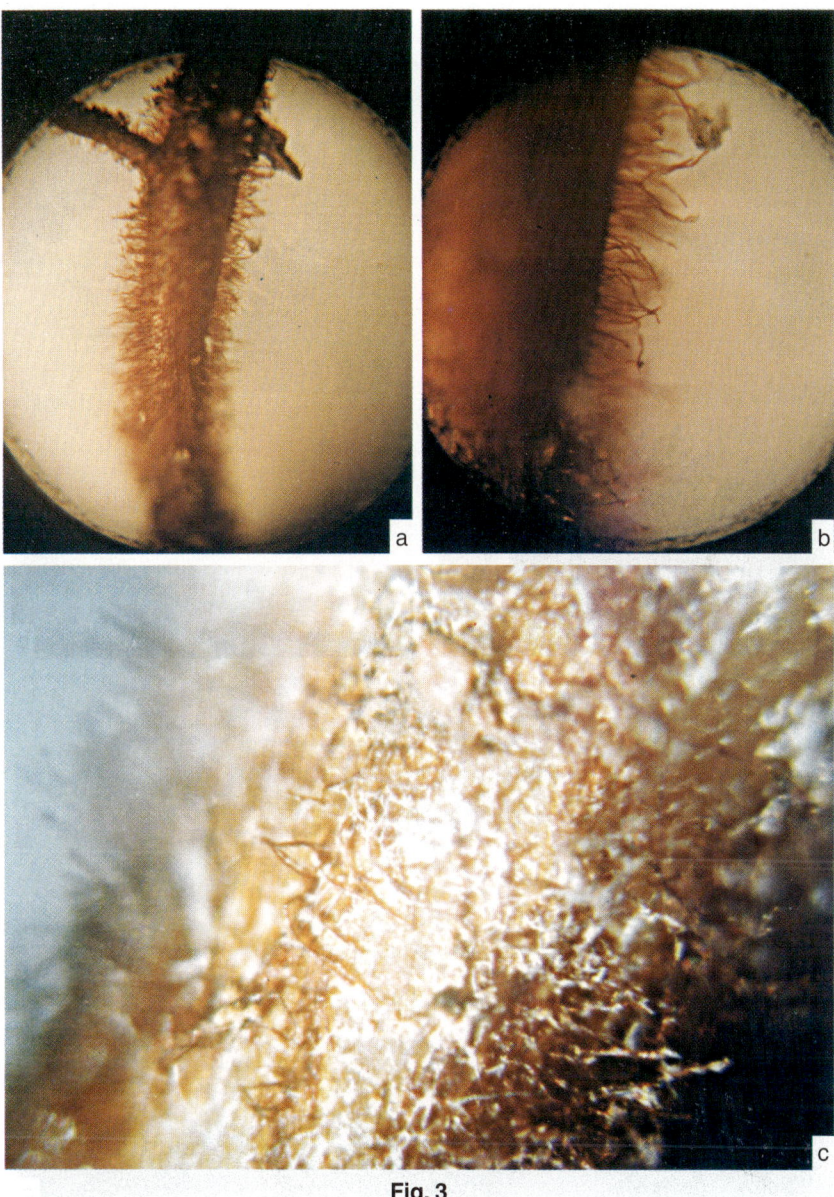

Fig. 3

a. Reddish persistent root hairs throughout the root system in Cashewnut tree (*Anacardium occidentale*) (x 75)

b. A portion of mature root with root hairs in cashewnut tree (x 200)

c. Ectomycorrhizal and root hair association in cashewnut tree when grown in a soil with better aerobic condition (x 320)

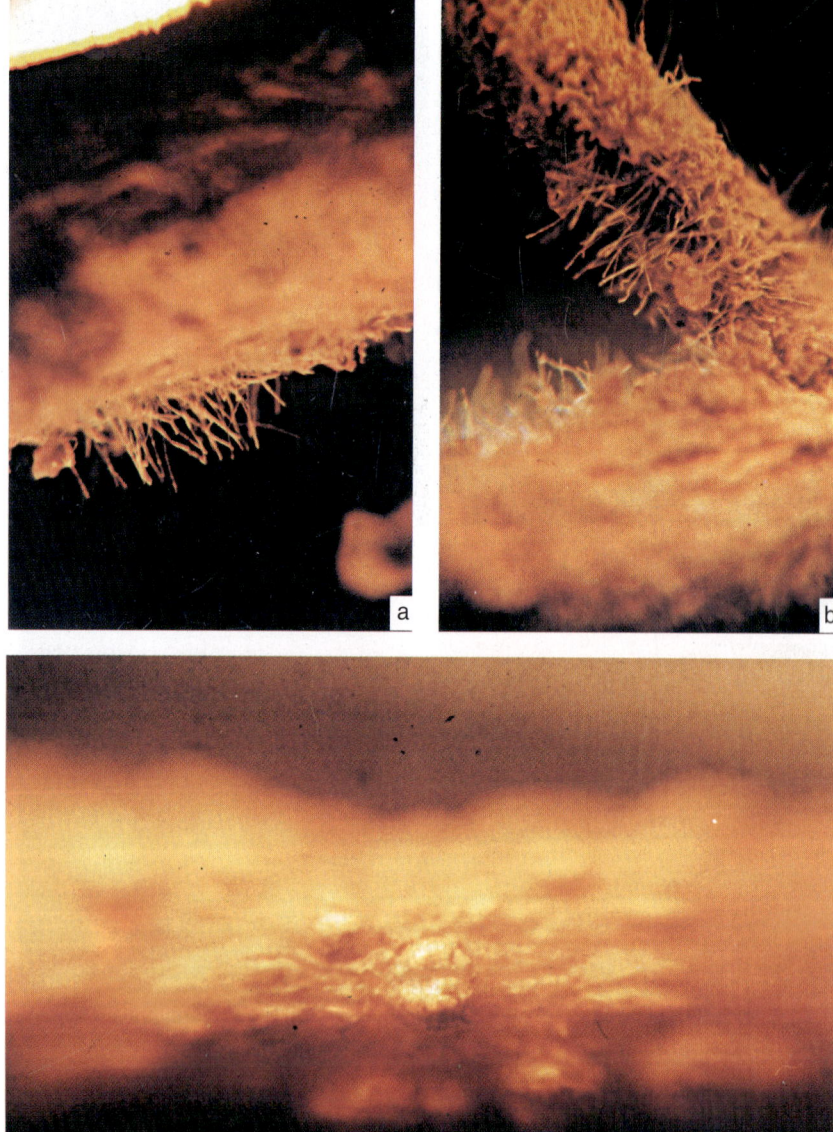

Fig. 4

a-b. Mature roots covered with persistent live root hairs in *Ailanthus malabarica* (x 275)

c. Mature roots of *Ailanthus excelsa* with prominent lenticels but without persistent root hairs (x 275)

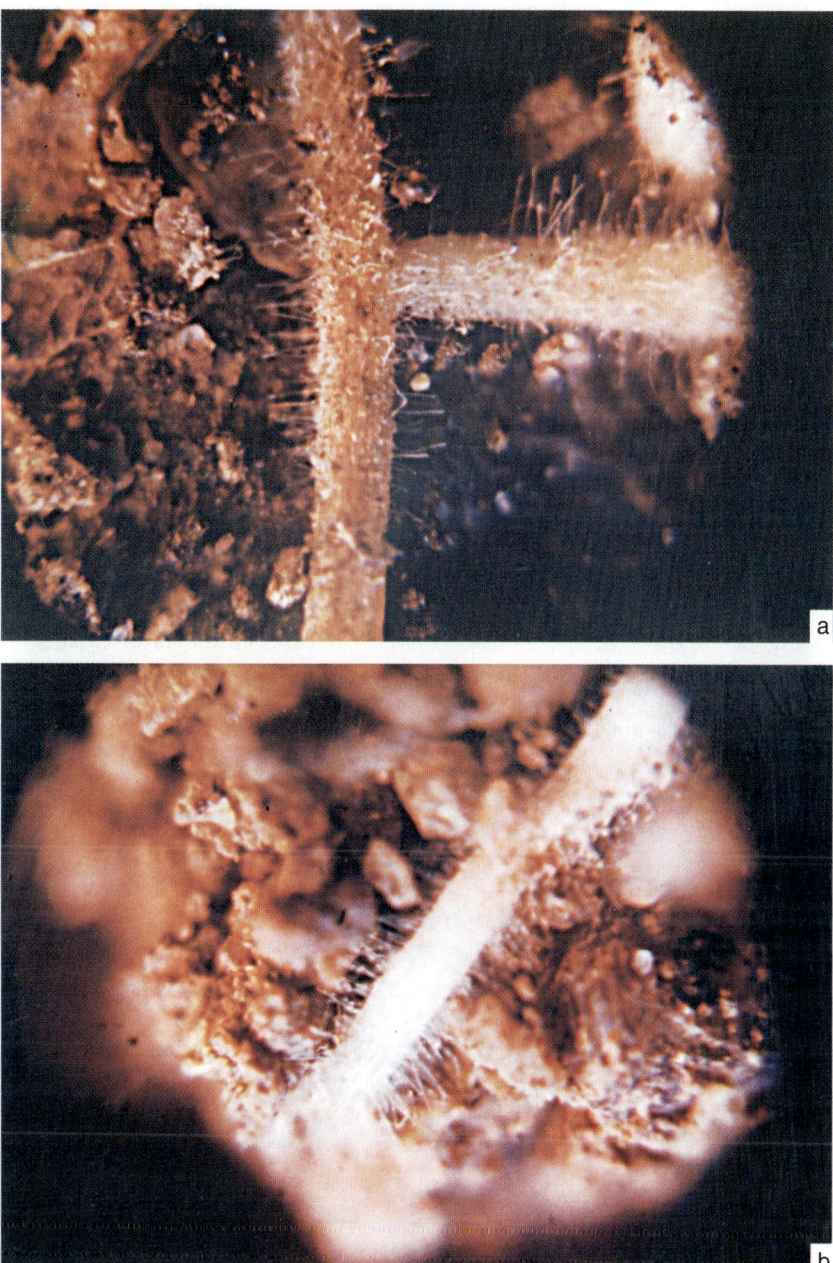

Fig. 5

a-b. Dense and longer root hair development in soil voids in *Ailanthus malabarica* (x 125)

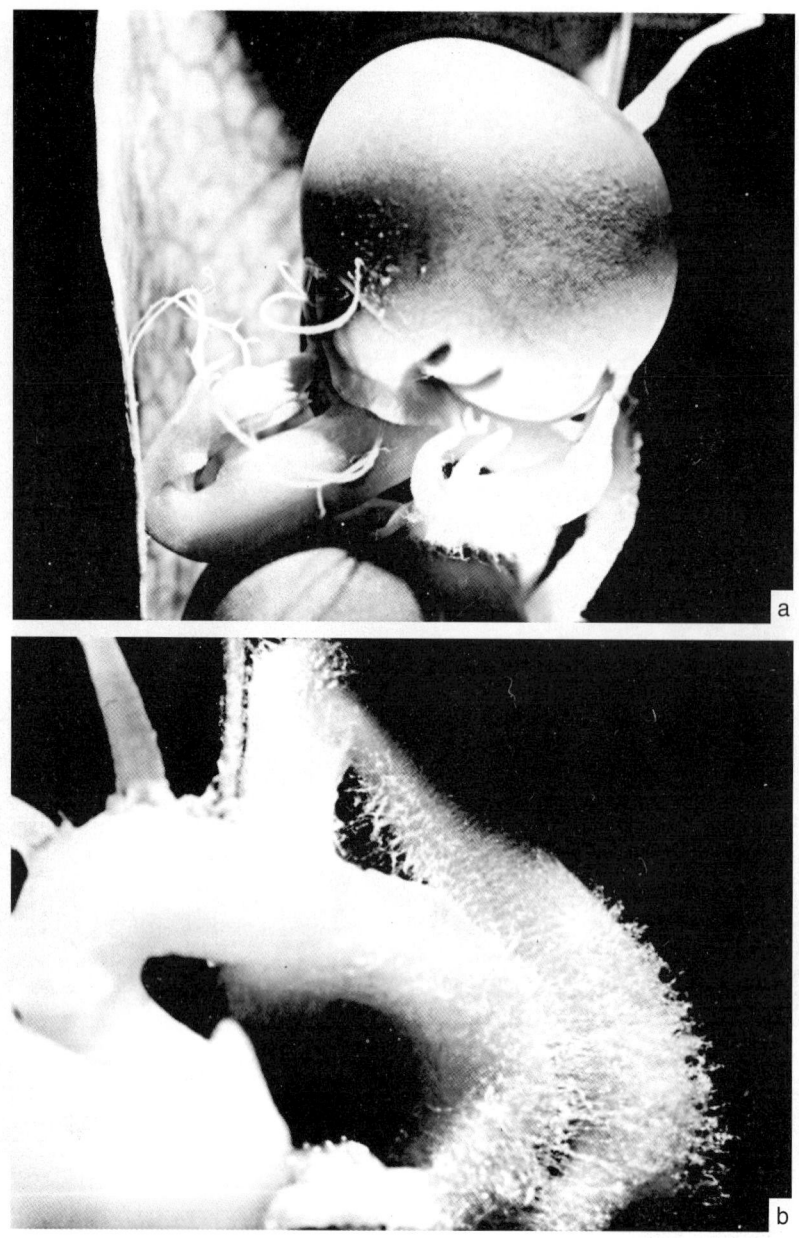

Fig. 6

a. Viviparous germination in *Pisum sativum* L. - pod and germinating seed with root hairs in air (x 10)

b. A portion of root showing dense root hairs in air in *Pisum sativum* L. (x 20)

Fig. 7

Effect of flooding on root hair development.

a & c. Poorly germinated finger millet (*Eleusine coracana*) seeds and radicle without root hairs under flooded condition

b & d. Normal germinated seeds with dense root hairs on radicles under aerobic condition in *Eleusine coracana*

e. Poorly germinated paddy (*Oryza sativa*) seeds without root hair development under flooded condition

f. Normal germinated paddy seeds with well developed root hairs on radicles under aerobic condition

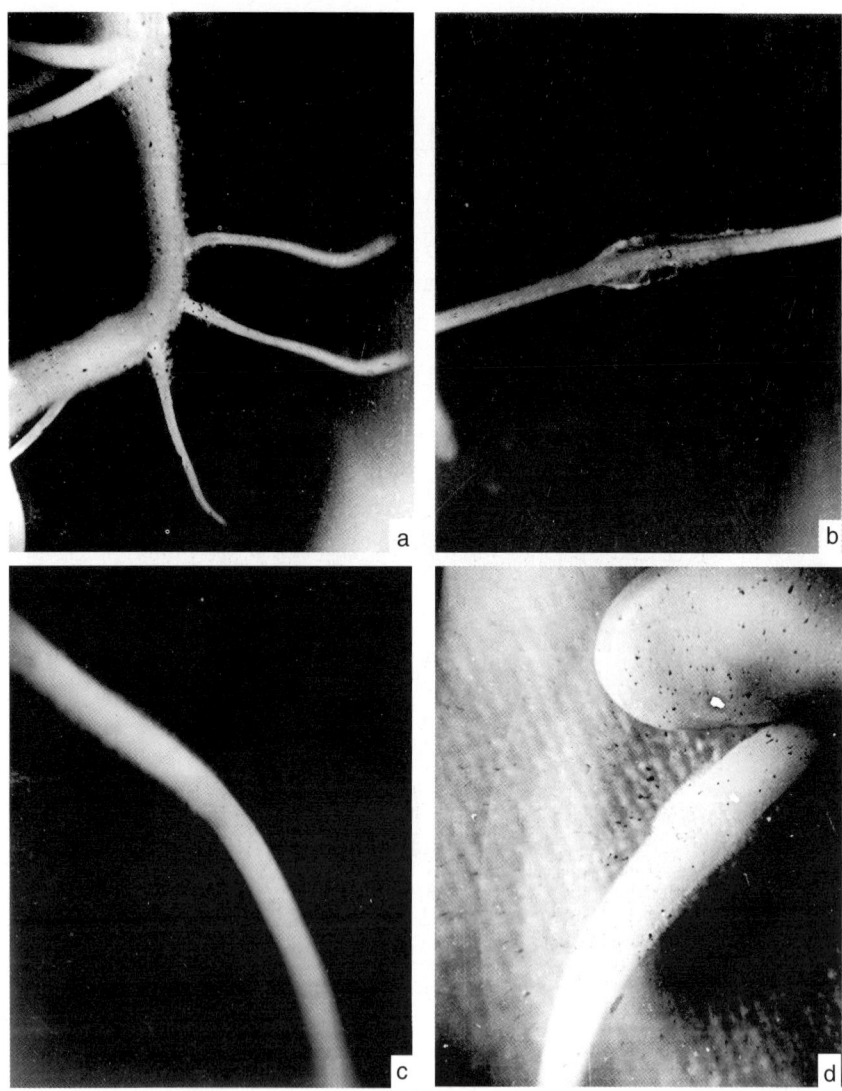

Fig. 8.

a. Dense root hair development on all the exposed roots and laterals in soybean (*Glycine max*) (x 33)

b. Root hair zone holding film of water after dipping in water in soybean (X 50)

c. Disappearance of root hairs within 2-3 hrs after a drop of water was put on root hair zone in soybean (x 40)

d. Parts of radicle in contact with water do not develop root hairs in soybean (x 35)

Fig. 9

Development of adventitious roots with dense root hairs exposed to air when tomato crop was flooded (*Lycopersicum esculentum*)

Fig. 10

a. Pot culture experiment showing effect of variation in aeration (oxygen enrichment) on growth of paddy (T_1= Control, T_2=Humus soil from 13 year old *Acacia nilotica* plantation, T_3 = leaf liter from *Acacia nilotica* plantation)

b. Experiment showing effect of reduced levels of irrigation (increased O2 enrichment) on yield of paddy

Fig. 11
a. Root hairs are normally well developed on the exposed roots in the tissue cultured *Vanda* (orchid) plants while roots inside medium lack root hairs (Photo by Dr. Vasantha Kumar Thimakapura)

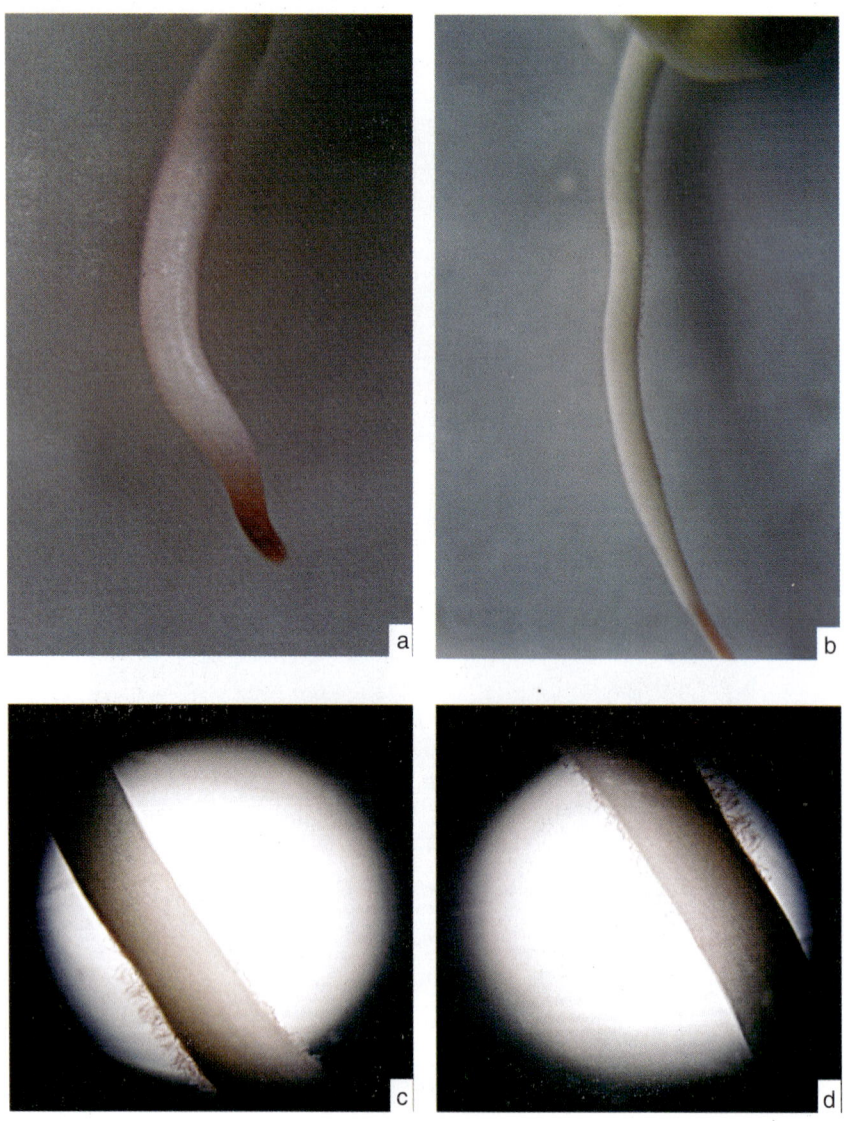

Fig. 12

a to b. Root tip turn red first and followed very soon by root hairs when 3-day old seedings of *Pisum sativum* are immerse in 1%TTZC soln.

c to d. Enlarged portion of root showing red stained root hairs with TTZC treatment.

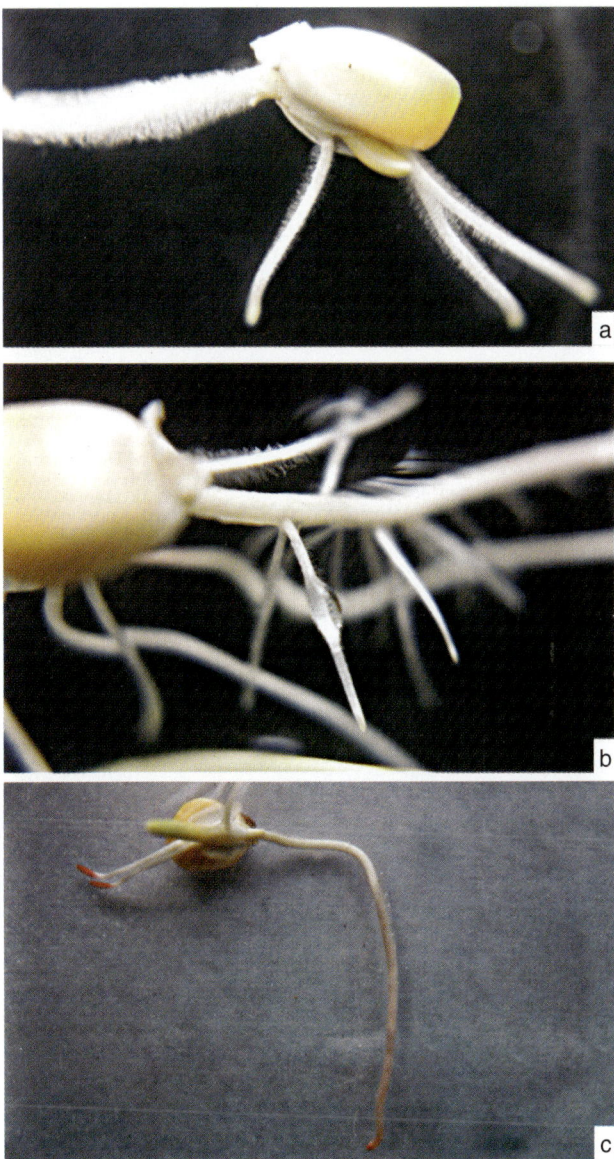

Fig. 13

a. Dense root hairs occurring on primary roots in maize *(Zea mays)*

b. Droplet of water remaing unabsorbed in the root hair zone even after 4 hours. This disproves water absorptive role of root hairs. Also observe abundant adventitous roots produced as a result of submersion of some roots

c. Root tips and root hair in maize have turned red with TTZC treatment

Index

2, 4-D 33, 35
ABA 84
Acacia nilotica 90, 91
Acer rubrum 102
Acidonia 11
Acorus calamus 161
Actinorhizal plants 46, 62, 63, 87
Active uptake 76
ADP 73, 89, 106
Adventitious roots 13, 55, 57, 58, 68, 75, 101, 121
Aerated medium 25
Aerenchyma 57, 67, 73, 80, 101, 103, 104, 134
Aerial root hairs 28, 118, 119
Aerial roots 7, 26
Aerobic environment 65
Aeroponics 68, 124
Aero-tropism 102
Agalinis purpurea 51, 129
Agrobacterium rhizogenes 35, 118, 153
Agropyron cristatum 10
Ailanthus malabarica 8, 10, 11, 18, 60, 113, 117
Air lacunae 78
Alfalfa 42
Alnus 45, 87
  glutinosa 46, 142
  rubra 131
Alternate furrow irrigation 108

Alternathera sessilis 9, 26, 27, 81, 118, 122
Amaranthus torreyi 13
Amaryllidaceae 8
Anacardium occidentale 5, 10, 11, 18, 113, 117
Anacharis (Elodea) 137
Anaerobic environment 107
Anaerobic respiration 72
Anaerobic stress 58, 164
Anaerobiosis 55, 57, 73
Annonaceae 6
Anoxia 67, 74, 89, 100, 106, 137
Aquatic plants 6, 24, 25, 48, 49, 50, 79, 113
Aquatic root hairs 28, 118, 122
Arabidopsis thaliana 31, 33, 134, 155, 159, 166
Arachis hypogea 6, 8, 13, 41, 64, 123, 134, 152-153
Araucariaceae 6
Air lacunae 78
Aristida pungens 54
Armoracia lapathifolia 35
Asteraceae 10
ATP 58, 65, 67, 71, 72, 73, 86, 87, 120
ATP/ADP ratio 104, 139
Auxin 33
Avena 38
  sativa 127

Avicennia 79, 81
   marina 137
Azospirillum 42, 46, 132, 133, 163
   brasilense 14, 32, 35, 133

Banana 63
Barley 34, 114
   cv. Salka 34, 40, 114
   cv. Zita 34, 40, 114
Beta vulgaris 69
Betula 98
Betulaceae 47
Biotechnology 34
Black gum 48
BOD 27, 81
Branched root hair 13
Brassica spp. 53
   napus 165
   oleracea 165
   rapa 165

Callitriche 78
Callose 45
Callosic plug 46
Cashewnut 63
Cassava 64
Casuarina 88, 134
   equisetifolia 6
Cation exchange capacity 52
Cell wall 15, 16, 18, 19, 29, 34, 42, 44, 45, 46, 58, 72, 82, 118, 128, 140, 152
Cellulase 34
Ceratonia siliqua 83, 135
Cercis canadensis 10
Chamaecyparis obtuse 152
Chicory 52
Chinese cabbage 52
Cinnamomum 6
Cisternae 15
Citrus 75
Cladium mariscus 136
Clayey soil 66
CO_2 concentration 98, 99
CO_2 evolution 97

Comptonia 134
Coprosma robusta 39
Corn 99, 102
Cortex 57
Cowpea 100
Cristae 119
Crop growth 84
Crop yield 84, 91, 92, 95, 97, 100, 109, 111, 114, 124
Cupressaceae 6
Cuticle 18, 19, 24, 56, 58, 77, 83, 113, 120
Cutin 19
Cuticulirization 24
Cyanotis cristata 49
Cyperus eleusinoides 6, 26, 49
Cytochemical assay 69, 163
Cytochrome oxidase 72, 119
Cytology 14
Cytoplasm 46, 71
Cytoplasmic organelles 14, 119
Cytoplasmic streaming 43, 44, 45, 73, 108, 120

Dictyosomes 15
Differentiation 24
Dimorphic root hairs 14, 117
Dipterocarpaceae 47
DNA 15, 16, 22
Drip irrigation 92, 107, 108, 124
Drought resistance 53
Dry soils 65

Ectomycorhizal relationship 50, 90, 115, 123
Ectomycorrhizae 47, 51
Ectomycorrhizal infection 50
Eichhornia 26
   crassipes 6, 15, 25, 26, 27, 49
Elaeagnus 45
Eleusine corocana 6, 70, 88
Elodea 24, 25
   canadensis 17, 61, 138, 18, 136
Endodermis 61
Endomitosis 15

Endoplasmic reticulum 14, 15, 46, 73
Endopolyploidy 16, 118
Epidermis 24
Epiphytic vascular plants 50
Equisetum 11, 143, 152
   hyemale 140
Ericales 6, 51
Eriocaulon decangulare 156
Erwinia carotovora 52, 139
Ethanol 72, 100, 103, 105
Ethylene 33, 103, 156
Ethylene synthesis 67
Eucalypts 50, 89, 135
Eucalyptus
   camaldulensis 102, 160
   globulus 160
   grandis 75, 101, 136
   hybrid 89
   robusta 101, 136
   saligna 101, 136
   teretecornis 89

Fagaceae 47
Feeder roots 11
Ferocactus acenthodes 53, 154
Finger millet 65, 89
Flavonoides 42
Flood irrigation 110
Flood tolerant species 101
Flooding 75, 102, 103, 106
   Effect of 75
Frankia 45, 46, 62, 88
Fraxinus pennsylvanica 159
Furrow irrigation 92, 108

Galium verum 50
Gaseous exchange 65
Genetic control 33
Geranium 127
Gleditsia 5, 60, 151,
   triacanthos 8, 10, 11, 18, 41, 117
Gloriosa 83
   superba 138
Glyceria maxima 161

Glycine max 34, 85, 88, 165
   var. Acme 34
   var. Harosoy 34
Glycolysis 71
Gnephalium 142
Golgi bodies 14, 46
Golgi vesicles 15
Graminoid roots 47, 48
Granite soil 65
Grasses 8, 21, 22, 37, 53
Ground nut 34
Growth 32
Growth respiration 106, 115
Gymnocladus dioica 10
Gymnosperms 47

Hairy roots 63
Halophytes 77
Haustoria 14, 51, 52, 129
Helianthus annuus 83, 84, 122, 141, 156, 162
Heteranthera zesterifolia 25
Histochemistry 43, 71
Histochemical 71
Hordeum 152
   vulgare 16, 56, 152, 165
Hormones 33
Hydrocharis 22, 137
Hydrophytes 24, 25, 29, 39, 54, 58, 68, 73, 76, 77, 78, 81, 118, 122
Hydroponics 32, 33, 69, 124
Hyoscyamus muticus 65
Hypathorine 50, 51
Hypodermis 57, 60
Hypolaena fastigiata 9
Hypoxia 67, 76, 100, 105
Hyrdilla verticillata 49

IAA 33
Ion exchange 39
Ion uptake 56, 59, 166
Iphegenia 83
   indica 138

Jussiau repens 26

K+ influx   40
Kalanchoe fedtschenkoi   7, 13, 68, 155
Kinetin   35
Kingia australis   63
Kreb's cycle   70
Lauraceae   6
Lectins   44
Leghaemoglobin   87, 123
Legume root hairs   16
Leguminous plants   41
Lenticels   81, 103, 119
Lepidium sativum   11, 15
Lettuce   39
Leucodendron   9
Leuminous plants   41
Liliaceae   8
Lipo polysaccharide   43
Lipochito-oligosaccharides   43
Lipo-oligosaccharides   42
Liquidambar styraciflua (sweet gum)   60
Liquification   60
Liriodendra tulipifera (yellow poplar)   60
Littorella uniflora   6, 135, 145
Lobelia dortmanna   6, 7
Lolium perenne (rye grass)   75, 138
Lotus corniculatus   18, 34, 156
Lucerne   47
Ludwigia   26, 27, 80
  adscendens   26
  octovalis   26, 27, 68
  peploides   79, 140
  repens   68
Lupin   32
  angustifolius   162
  luteus   149
Lycopersicum esculentum   35, 85

Magnoliales   47, 117
Maize   31, 74, 89, 100, 102, 104
Medicago
  sativa   32, 42, 47, 148, 152, 165
  truncatula   23, 45, 160

Melaleuca quinquenervia   75, 101, 160
Menespermaceae   8
Menyanthes trifoliata   161, 136
Meristematic region   60
Micro fibrils   19, 23
Micro tubules   16, 17, 23
Microspectrophotometry   16
Mineral nutrients   30
Mineral uptake   37, 39, 47, 56, 59, 76
Mitochondria   14, 15, 46, 58, 72, 73, 119, 120
Mitochondrial ribosomes   16
Molina coerulea   50
Mucigel   53, 62
Mucilage   54
Mucilaginous layer   32
Mud roots   26
Mutant   34
Mycorrhiza   47, 48, 63
Mycorrhizal colonization   95
Mycorrhizal infection   47, 49, 50, 100
Mycorrhizal plants   62, 98
Mycotrophy   47
Myricaceae   134
Myrica   134
Myriophyllum   78
Myriophyllum triphyllum   6, 49
Myrtaceae   47

N_2 – fixation   87
N_2 fixing plants   107
Nematodes   52
Nicotiana tabaccum   69
Nitrogen   32
Nitrogen fixation   123, 130
Nitrogen fixing microorganisms   41, 62, 86, 110, 123
Nitrogenase   87
Nod factor   42, 43, 45, 46, 110
Nodulation   41, 42, 43, 45, 47, 85, 92, 107, 128, 133
Nodule   87
Nodule formation   41, 42, 47

Nucleohistone 15
Nucleoli 15, 22
Nucleus 14, 23, 42
Nuphar advenum 148
Nutrient uptake 37, 48, 63, 109, 115, 119, 124
Nyassa sylvatica 48, 144

O_2 concentration 79
O_2 stress 66, 110, 113, 114
Opuntia 5
   ficus – indica 53, 54
Orange tree 65
Orchidaceae 8
Orchids 51
Organic farming 95
Ornithopus 84, 122
Orobanche ramosa 52, 143
Oryza sativa 70, 88, 146, 154
Oxygen concentration 64, 128
Oxygen deficiency 84, 89, 102, 105, 107
Oxygen diffusion 57, 71, 103, 105, 109, 113, 115, 121, 123, 124, 128, 150, 162
Oxygen enrichment 90, 124
Oxygen intake 51, 115
Oxygen supply 55, 67, 106
Oxygen transport 48, 104, 120, 158

Paddy 90
Paddy soil 65
Pandanus 7
Pathogens 52
Pea 32, 71, 88, 89, 103, 109
Pectolyase 34
Penicillium bilaii 32, 109, 118, 143
Perneltya macrostigma 133
Persistent root haris 11, 113, 117
Persoonia 11
Pest resistance 114
pH 15, 32, 47, 131, 141, 162
Phalaris arundinacea 161
Phaseolus 69
   vulgaris 67, 108

Phosphorus uptake 39, 40, 114, 115, 162
Photosynthesis 18, 19, 24, 79, 99, 119
Phragmites communis 161
Physiological dryness 77
Picea glauca 146
Pinaceae 21, 47
Pinus
   contorta 102
   echinata (short leaf pine) 60
   edulis 18
   elliottii 97
   lambertiana 131
   serotina (pond pine) 75
   sylvestris 98
Pisum 122
   sativum 18, 28, 53, 67, 70, 73, 83, 88, 144, 149
Pistia stratioites 6, 25, 78
Plant growth 106, 108
Plant response 100
Plasma membrane 15, 16, 18, 34, 44, 140
Plasmadesmata 61, 72, 86, 87, 120, 148
Plasmalemma 23
Plasmodiophara brassicae 53, 165
Plastids 15, 46
Pneumatophore (air roots) 81
Poaceae 48
Polyalthia longifolia 6
Polycephalum 142
Polyploid 15, 16, 22
Populus euroamericana 48
Pores 72, 120
Porous soil 66, 91
Potato 107, 129
Potometer 60
Pratylenchus mediterraneus 154
Proteaceae 11, 63
Proteoid roots 63
Protoplasm 14, 59, 60, 119
Protoplast 34

Prunus persica 71, 132
Pseudomonas spp. 46
Pseudotsuga menziesii 21, 32
Pterocarpus officinalis 7, 86, 159
Pycnonia 11

Quercus borealis 5

Radicle 66, 67, 68
Raphanus 83
  sativus 17, 61, 138
Repiratory gills 55
Respiration 55, 92, 134, 140
Respiratory quotient 56
Respiratory role 63
Restricted irrigation 90, 109
Rhicadhesin 45
Rhizobium 41, 42, 43, 44, 45, 47, 70, 86, 123, 130, 134, 154, 161
  leguminosarum 44, 53, 139, 142
  meliloti 42, 43, 44, 128, 165
  trifolii 43, 44, 85, 166
Rhizophora 81
Rice 48, 55, 65, 67, 128
Ribosomes 14, 15
RNA 15, 22
Root hair
  Abundance 50
  Density 24, 30, 39
  Development 21, 24, 30, 32, 33, 65, 66, 84, 85, 92, 107, 109, 114, 115, 120, 136
  Elongation 50, 51
  Growth 39, 95
  Length 33, 34, 38, 40, 50, 58, 109, 134
  Life span 10
  Morphology 41, 108, 127
  Number 8, 34, 40
  Occurrence 5, 50
  Respiration 95, 98, 107, 108, 149
  Infection 131
  Management 97, 106
  Size 8
  Structure 13
  Surface 54
  Surface area 9
  Tip 38, 41, 43, 44, 52, 59, 119
  Vacuole 38
Root hair activity 77
Root hair infection 45
Root hair zone 8, 32, 52, 53, 58, 65, 71, 74, 84, 88, 98, 121, 122, 123
Root 55
Root aeration 76, 128, 129, 137
Root anaerobiosis 135, 140
Root culture 65
Root epidermis 49, 62
Root growth 99
Root meristem 56, 69
Root nodule 41, 62, 69
Root parasitism 51
Root porocity 104,
Root respiration 71, 72, 97, 98, 106, 110, 132
Root system 55
Root tip 60, 70, 71
Russian thistle 39

Salicaceae 47
Salinity 32
Salix
  alba 103
  caprea 162
  cinerea 162
Salt respiration 56
Sandal 52
Santalum album 7, 51, 148, 154, 164
Schaenoplectus lacustris 152, 161
Scrophulariaceae 14, 52, 117, 129
Secale sereale 5, 8, 9, 139
Seed germination 65
Seed yield 108
Sinapsis alba 15, 141
Soil aeration 66, 85, 92, 95, 97, 104, 109, 114, 115, 124, 139, 161
Soil anaerobiosis 97, 100, 105, 115, 121
Soil compaction 107, 108, 127

Soil environment 109
Soil management 84
Soil moisture 30, 74, 92, 108, 114, 150
Soil oxygen 149
Soil porocity 99, 114
Soil voids 58, 65
Soil working 114
Solanum tuberosum 64, 118
Soybean 35, 48, 85, 86, 88, 89, 99, 119, 124, 130
Spathoglottis plicata 50, 148
Stomata 56, 58, 77, 83, 84, 119, 122, 123, 162
Stress roots 75
Suberin 19
Suberization 60, 61
Sunflower 102, 108
Surface irrigation 95
Symbiosis 46
Syzygium cumini 19

Tagetus patula 13
Tamarindus indica 6
Taxodiaceae 6
Taxodium disticum 102
Teak 30, 66, 90, 99, 165
Tectona grandis 50, 90, 99
Terrestrial plants 25, 27, 50, 58, 76, 82, 113, 118
Thielaviopsis basicola 52
Tinospora cordifolia 8
Tip callose 19
Tip growth 22
Tissue culture 65, 123
Tissue culture medium 63, 123
Tobacco 99, 143
Tomato 28, 34, 39, 64, 66, 85, 121, 129
Tradescantia 16, 152
Transpiration stream 105
Trianea bogotensis 40, 61

Trichoblasts 15, 22, 23
Trifolium 69
  repens 43, 133
  sub-terraneum 14, 41, 85, 166
Triticale 154
Triticum 31
  aestivum 128, 140, 146, 165
Typha latifolia 161

Ultra-structure 14, 71
Urtica dioica 17, 128

Vacuolar acidity 106
Vacuoles 15, 72
VAM 26, 47, 48, 86, 115, 123
VAM colonization 50
VAM hyphae 49
VAM infection 49, 63
Vanda 63
Vascular plants 6
Vicia 19
  sativa 52, 144, 154
  faba 55, 60, 140

Water absorpotion 37, 38, 56, 119, 133
Water logging 100, 105, 106, 128, 132, 155
Water management 84
Water uptake 37, 59, 63, 76
Water uptake efficiency 38
Water use 109
Wet soils 58, 65, 102
Wheat 10, 32, 39, 40, 46, 98, 103, 105, 128
  cv. Kraka 10
  cv. Kosack 10
  cv. Foreman

Xerophytes 17
Xylem elements 60

Zea mays 56, 61, 70, 74, 89, 129, 134, 135, 139, 145, 151, 153